ON THE TRAIL OF WOLVES

ON THE TRAIL
OF WOLVES

Philippa Forrester

BLOOMSBURY WILDLIFE
LONDON · OXFORD · NEW YORK · NEW DELHI · SYDNEY

To my Mum who taught me to be bigger and braver but to always try to understand rather than judge.
&
To all women who are trying to fit themselves and their work around loving everyone else.

BLOOMSBURY WILDLIFE
Bloomsbury Publishing Plc
50 Bedford Square, London, WC1B 3DP, UK

BLOOMSBURY, BLOOMSBURY WILDLIFE and the Diana logo
are trademarks of Bloomsbury Publishing Plc

First published in Great Britain 2020

A catalogue record for this book is available from the British Library

Library of Congress Cataloguing-in-Publication data has been applied for.

ISBN: HB: 978-1-4729-7204-0; ePub: 978-1-4729-7203-3;
ePDF: 978-1-4729-7206-4

2 4 6 8 10 9 7 5 3 1

Map by Gus Hamilton James

Typeset in Bembo Std by Deanta Global Publishing Services, Chennai, India
Printed and bound in Great Britain by CPI Group (UK) Ltd, Croydon CR0 4YY

To find out more about our authors and books visit www.bloomsbury.com
and sign up for our newsletters

Contents

The Greater Yellowstone Ecosystem

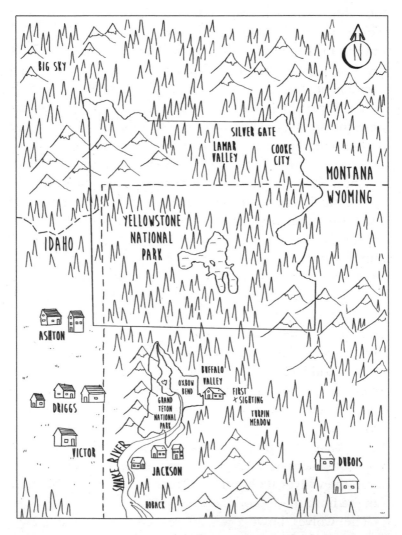

Although not perfectly to scale, this map gives you a gist of the lie of the land. Dashed lines are state lines, while the solid line indicates the boundaries of Yellowstone National Park.

Prologue

I hear the wolves while I'm working alone in our snow-clad log cabin in the wilds of Wyoming. I have never heard wild wolves before, but I am in no doubt about who it is calling outside in the snowy valley. I stand on the deck and wrap my cardigan around me, shivering, both with cold and excitement.

It is surprisingly easy to track their progress – not just from the howling. The still air is full of them, alive, as if they have a static presence. They're travelling towards me, along the river that runs just a few hundred yards from our house.

I hear dogs too, in a cabin further downriver. They are apoplectic. I wonder if they could be the golden retrievers I saw the other day. I haven't seen any sign of Dave, who apparently lives in that cabin, so I haven't been able to ask him about them. Anyway, their best 'guard dog' barks seem a bit pointless – spiky rather than intimidating, and, I would imagine, rather irritating to a wolf.

A group of three Canada geese fly up from the river, honking in tuneless alarm as they try to arrange themselves into the correct flying formation. Are the wolves right there?

Two paddocks away, in the branches of a grey-green aspen tree, crows and magpies gather together in a large group. They caw loudly and peer in the direction of the river. I have never seen them gather together like that. But I have heard that they will follow a wolf pack in the vague hope of snatching a piece of meat from a kill. Does that mean the wolves are hunting? Oh, why can't I see what they can see from the top of that tree?

The wolves howl again, closer now. A moment later comes a reply from way over the forested hill. A single howl. A lonesome wolf?

The call is so low in pitch that somehow it fills my heart, swilling around inside, like thick red wine in an over-large glass. Humans can only howl like that when it comes from deep inside our hearts. I know, I have done it – only once, with our dog. Funny, yet it was desperately sad at the same time – my heart was broken.

I rush upstairs to press my binoculars against the window... Nothing. Racing back down the slippery wooden steps to the ground-level windows, I peer between our neighbours' house and the trees... Still nothing. Upstairs again, two at a time, to squint at the hills rising from the river. I still can't see any wolves, but I can hear they are really close. So I return to the deck and just listen. After all, that is where the magic is.

Perhaps today I don't need to see them. Instead, I can imagine what the wolves are doing: they are travelling away from me now, a pack running swiftly together through the valley towards the forested hill where the lone wolf calls. They are organising each other, making a plan, planning to meet.

After a while, the sounds become fainter. Perhaps I should return inside, carry on writing. After all, you never catch a bunch of wolves sitting around in the woods procrastinating – they might howl a bit, but ultimately they just get on with what they need to get on with. I wonder what that is?

When I return to stand outside ten minutes later, the wolves are gone. I know it – not just because their sounds are gone but because I can feel it in the air. Every other creature is quiet, and the air is clear – the wolves' static gone. But those calls linger in my head.

Whether they are meant this way or not, these calls are a call to me – a wake-up call, adding fuel to a smouldering fire inside me. They make me realise where my passion lies, where my curiosity ignites: with the wolves in the wilderness.

CHAPTER ONE

The Phantom Springs pack

We have brought our family on their biggest adventure, from the Austenesque world of Bath in England to the Wild West. We are on assignment for *National Geographic* magazine, for a special one-off edition about Yellowstone National Park. Always ready for excitement, we packed up our three kids, ten bags (much of which was Lego), and a truck-load of camera kit, said goodbye to family, friends and pets (prompting many tears), and headed out into the wilds of Wyoming for a year. We are, for now, living in a log cabin where the nearest pint of milk for sale is an hour and twenty-five minutes away. The kids are in a school with only thirteen other children, and life is pretty much unrecognisable apart from the laundry and the dishes, which are sadly persistent wherever we go.

We are here to find and photograph wildlife, and we are all agreed that this is one of the most exciting things we have ever done.

On our first weekend, we wrap up the kids and hoist cheap, plastic toboggans up a nearby hill. Ours are the only tracks in the snow. It feels odd not to be surrounded by people. In England, when it snows, any hill is immediately invaded by brightly dressed people slightly delirious with the novelty of it all. Not so here. There is no one else for miles.

After a few screeching, bum-smacking trips on the toboggan, I collapse breathlessly in the snow at the top of the hill and gaze at the view. From here, I can see mountains all around us — blue, grey and white — and the ribbon of steely-grey river as it leaves the white and green of the conifer forest and winds through the trees towards our little cabin. On either bank I can see small tracks, which might well have been made by beavers or otters. I trace their lines upriver and then, quite far off, I spot a moose with a calf.

Moose are most odd, but I love them for it. They are dark velvet-brown with long, long legs that look ridiculous but are probably useful when the snow is this deep. In the spring mating season, the males have huge, broad antlers. I read once that in rare situations a cow moose can grow antlers too, but that is usually down to a hormonal imbalance, which made me chuckle and feel rather good about myself: until that point I had thought my own hormonal imbalances had weird consequences, but at least I don't end up growing antlers.

In any case, male or female, their most noticeable feature is the shape of their heads, which are rectangular with an unimpressive, disproportionately small ear at each of the top two corners. Between the bottom two corners droops a wobbly, over-large upper lip that covers the lower lip completely, making them look like both a blockhead and a goofball at the same time. That said, their huge size means they retain an aura of majesty and do command respect.

I'm pleased they aren't any closer. Apparently, a mother moose with a calf can be about as dangerous as a grizzly bear, especially if the mother feels threatened – and more especially if you accidentally find yourself between that mother and her baby. Our free-fall tobogganing kids could have found themselves sliding into that position without realising it.

Mum and baby are standing in the middle of the water. I can't help but wonder at how uncomfortably cold that must be, but moose like water. They are known swimmers, and the water often provides them with an escape route from predators.

The mother lifts her head high, suddenly alert. Perhaps she has heard something. She focuses her small ears on the forest. After a brief pause, she starts moving down the river towards our toboggan hill – slowly at first, waiting for her calf to get the message and follow, but then faster and with a sense of urgency. I wonder why. We are far too far away to bother her, and she wouldn't run towards us if we were.

She breaks into a lolloping canter, her long legs and large hooves kicking up the water, splashing her calf who is trying to keep up but, on little legs, is far slower.

From my vantage point, I am excited and nervous at the same time. The mother has clearly detected a predator, which at any other time of year could be a bear, but bears are still hibernating, so it is more likely a wolf.

Behind them, a dark shape leaves the tree line; clearly on two legs, it is a man. How disappointing. He can't see the mother and calf, as they have rounded the bend in the river. He wears snowshoes and walks quickly along the bank. He seems to be looking for something.

As he approaches the base of our hill he hears the kids and, looking up, spots me and waves. I wave back and, with a lingering look at the moose family disappearing into the forest, I set off to meet him.

He is blond and super-tanned, much older than us but almost certainly fitter judging by the speed he can go on those snowshoes.

'Hi!' I call as I approach. 'You OK?'

'Hi!' he says. 'You must be the British family who moved in up the road?'

So we do have neighbours.

'Yes, that's us.' I'm guessing the accent gives it away. 'I'm Philippa and this is Charlie.'

'Sorry?'

'Philippa.' I'm getting used to this now. It seems no one in America has ever heard the name.

'Fileepa,' he says uncertainly, feeling it out. Then, 'What a lovely name.' They all say that too. Americans, in my experience, are usually very concerned not to cause offence and are in fact extremely welcoming.

'Thanks. Where do you guys live?'

'Just a little further down the valley, right near the river.'

David, it turns out, has lost his dogs.

'Have you seen them? Two goldens. They went out to pee earlier and never came back in.'

'Golden retrievers? We haven't, but we can keep an eye out for them.'

'It isn't like them to run off. Sometimes they can get on a smell and then they go deaf for a while. Then it doesn't matter how much you call them, but they've been gone for hours now and that's not like them at all. Could be that the wolves have them.'

'Does that happen?' I ask.

Dave shrugs, squinting into the distance, the lines on his face clearly accustomed to that position, his blue eyes a match for the sky.

'Occasionally, what they do apparently is they'll come calling and kind of seduce a dog. The dogs will go up to them to play, and then the wolves just rip 'em apart. I know the wolves have been around here.'

I think of the wolves I heard the day before. Surely the wolves here have enough to eat. Why would they be interested in a couple of pet dogs?

We chat to Dave a little longer as we all watch the boys throw themselves down the hill. He is a realtor and – despite working in the nearest town of Jackson and having to drive for over an hour every day, often in blizzard conditions – he tells us he couldn't live there, 'not when it's so beautiful up here.'

I spend the rest of the day looking out for the dogs but see nothing. But that night, just as I am going to bed, I glance out the window and spot a pair of headlights bumping over the snow. It's Dave on a little quad bike, wearing a head torch. I watch its slim beam scanning back and forth. So the golden retrievers still aren't home.

'If they are alive, I don't see how they could survive out there tonight,' I murmur to Charlie. As I lie in bed, the missing dogs bother me.

Next day, reading the local paper in a coffee shop in town, I'm delighted to find a map that shows the territories of all the wolf packs in the Greater Yellowstone ecosystem. I'm quick to find us on that map. Ours is the Phantom Springs pack, currently made up of eleven wolves. According to the article accompanying the map, last year wolves in Wyoming killed forty-one cattle, thirty-three sheep, one goat and one dog. Two more might be added to that list this year, but I hope not.

Although wolves are protected in the nearby Grand Teton and Yellowstone National Parks, I'm surprised to find they are currently delisted in Wyoming, meaning they can legally be hunted as long as the hunter has a permit or if the wolves are predating local livestock. So last year, in exchange for the seventy-five animals they killed, thirty-three wolves lost their lives to 'predator management'. And that figure doesn't include those that were legally hunted as 'trophies'.

They actually describe it as 'harvesting' wolves. For all that I would struggle to come to terms with a wolf taking my dog, I find the term 'harvesting' sticks in my throat.

I remain desperate to see a wolf.

In fact, although I know we have at least a few human and lupine neighbours, they are all a little elusive. I have seen one lady, Tracy, several times at school. Actually, she's always there, but it turns out she is the cleaner, teaching assistant and lunch assistant. I'm delighted to discover that she also lives in our valley. So next time I see her, I take a deep breath and start chatting.

'We are in the red cabin with the green roof.'

'I know it,' she smiles.

'We should share school runs,' I say. 'I am happy to bring your son in – just say when.'

She smiles again but leaves it there.

'Any time you fancy a glass of wine, just come round.'

I give her my number. Again, the warm smile but not the commitment.

This happened on more than one occasion. 'Don't forget that glass of wine,' I would remind Tracy cheerfully as my children charged up and down the small corridor, looking for lost gloves and pulling on snow boots.

I remembered new people at our school in England. How often had I taken the time out of my busy life to make them welcome? Apart from a smile and a quick chat, I don't think I did.

One day my husband asked if her son might like to come round and play. Apparently, she hesitated before replying.

'Well … you see,' she said, 'it's just that we are all wrestlers.'

'Right, of course,' said Charlie before hustling the children into the car and driving home.

'What did she mean?' I ask in the kitchen as I clear up the mess of snow pants, boots, gloves and hats deposited by the kids at the back door.

'I don't know.'

'Are they all wrestlers?'

'I'm not sure.'

'Why does that mean he can't do a playdate?'

'I didn't ask.'

Why don't men ever gather enough information?

'So are they wrestling the whole time? Even at home?' I try to push it away, but I can't help imagining the scene. Inside a lonely cabin in the middle of all that snow, like a scene from *The Incredibles*, the whole family dressed up in wrestling gear. 'I wonder if they get all the cushions off the sofa and use them as crash mats.'

'Well, it's one way to keep warm, I guess.'

But I soon realise why Tracy doesn't have time to socialise: she's a bit like Miss Rabbit from *Peppa Pig*, turning up in all sorts of jobs, always smiling. As well as being the cleaner, she is also the librarian and assistant at school and – from the little I can gather – she also works on the family ranch, helps with the cabins that are rented out to tourists, works for the post office and the fire department and is evidently heavily involved in the wrestling world too. Obviously very busy.

Then, one day, she calls. I am particularly excited because, apart from my husband, this is the first person who has ever called me on my American 'cell'. Perhaps this will be it – the first glass of wine with my neighbour.

'I'm just calling to let you know that it is "Teacher Appreciation Week" next week.'

'Ah! Great! What is that?'

A pause. Then, 'It's a week where we let the teachers know how much we appreciate them.'

'Of course.' This, it turns out, is an American tradition I was ignorant of.

'So, on Monday I'll be…' Tracy reels off a list of kind deeds she will perform for the two teachers at school for every day of the week, ending with Friday 'when I'll be doing a special pulled pork lunch for both teachers and the kids. So I was wondering if you'd like to come to that and perhaps help?'

'Yes, I'd love to!'

Finally, I am part of something, practically a local.

'What can I do? I know, I'll do dessert.'

Both younger boys, happily drawing at the kitchen table until that moment, slam down their pencils to stare at me.

'What?' I ask innocently after saying goodbye to Tracy. But I know why they are staring. I have been banned from ever uttering the phrase 'I know, I'll do dessert.' Usually, if I forget, one of the family is there to remind me. They

politely inform the recipient of the offer that I have no
business with dessert and then they elaborate. They have
numerous stories to choose from, generally involving me
being late for dinner parties and turning up with runny
mascara and a Marks & Spencer roulade.

So I have really done it now. There is no Marks &
Spencer to rescue me here. I have no choice – I will have to
pull this one off.

And I do. With no internet, I fall to the library in town to
find a recipe book and end up making the most delicious
muffins – even though I do say so myself – like a true
pioneer woman.

Taking them along to the teachers' lunch also gives me a
chance to meet some other mothers who Tracy has asked to
bring in a dessert as well. I get chatting to one in particular,
Jen, who – between mouthfuls of Tracy's melting pulled
pork – tells me that Tracy is Mormon.

A penny drops.

'Mormons don't drink, do they?'

Jen shakes her head.

'So it was probably the wrong thing to do to invite her
for a glass of wine, then?'

Jen spits out her mouthful and we laugh a lot, and right
there and then I make my first friend.

CHAPTER TWO

How close does a bear have to be before you spray him?

It's funny transplanting a family – so much changes, yet so much stays the same. (As usual, there has already been a high level of divorce in the sock department, resulting in an already mountainous pile of single socks looking for partners.) But we soon fall into a new rhythm.

For Charlie, on a photography mission, the first job is to find beavers. In the absence of anyone else, I'm drafted in as first assistant. We decide to try the Buffalo Fork river nearest us first.

We take two cans of bear spray each. No sensible person goes hiking or off into the backcountry here without bear spray – a type of pepper spray that comes in a small can. Although the grizzly bears will almost certainly still be asleep, there is no telling when they might be starting to wake up.

There is a story – I don't know if it is accurate or the wilderness equivalent of an urban legend – that a Swedish woman, misunderstanding the point of bear spray, diligently sprayed her children with it before setting out on a hike. They all ended up in hospital. Bear spray is pretty powerful stuff only designed for spraying bears when they charge you.

'Be noisy, travel in a group and carry bear spray.' This is what we have been advised to do, over and over but especially by Steve Cain, head of wildlife at Grand Teton National Park. It's called being 'bear aware'. Most grizzly bear attacks occur when a bear has been taken by surprise, such as by one person hiking quietly, or when a mother bear is being protective of her cubs. Or, I gulp, in early spring when the bears have just woken up because they are very, very hungry.

So travelling with someone and making lots of noise will help you avoid an encounter with a grizzly. Should you end up in the unfortunate situation where a grizzly bear, not noticing that you are being 'bear aware', has decided to charge anyway, the idea is that you hold your ground until the bear gets to around twenty-five feet away then spray. The instructions on the can say, 'Aim at face and eyes of the bear' and 'Bear spray has been shown to reduce the length and severity of "maulings".'

I can't say I'm filled with the usual excitement about being in the wilderness, and being drafted in as first assistant because of bears, but Charlie deals with situations like this all the time, so I keep smiling as I step into the snowshoes and bend down to tighten them up.

I immediately learn my first lesson as I face-plant onto the cold snow. Mental note for future: 'Don't try to bend forwards on snowshoes, in snow, with a heavy rucksack full of kit on your back. It throws your balance right off.' Of course, my husband already knew that.

Snowshoes finally strapped on tight, it takes a huge effort to haul the heavy rucksack back onto my shoulders, but finally we are ready to go. Trying not to sink into the snow, we head into the woods.

I had no idea how difficult snowshoeing would be. Each step, under the weight of the kit, is an enormous effort. We don't talk, we don't have the spare lung capacity. We are still adjusting to living at altitude, let alone being active at altitude. I'm sweating after just fifteen paces. I have put on way too many layers but can't take them off because I'm not sure I'll be able to haul the rucksack back onto my shoulders again.

I stare into the trees and suddenly remember that we are meant to be noisy.

'Um, how far actually is twenty-five feet, then?' I shout.

Charlie glances up at me.

'What?' he says at normal volume.

'Twenty-five feet,' I repeat loudly. My voice has a definite tone of forced jolliness though, for the life of me, I can't work out why. Why would a surprised bear care if I was in a jolly mood or not before deciding whether to attack? 'I can't really tell how far that is,' I call.

Charlie pauses, panting, bent over by the weight on his back.

'Well, considering you once told the removals man that our bed was a six metre one, I'm not surprised.'

I ignore the jibe.

'It's just that if a bear were to attack...'

'A bear is not going to attack. A bear is not just going to run out of the woods and attack us.'

'But if it did, how close should it be before I spray the bear spray?'

'About from you to that tree there.'

Charlie puts his head down and starts trudging again. At six foot four, his legs are longer than mine, so he goes faster. My snowshoes feel so heavy, each step so exaggerated. Despite their large surface area, I still sink into the snow.

I regard the twenty-five-foot distance. I might well be dead of fright before a charging bear got that close to me; how would I be in my right mind enough to start using bear spray?

I need to keep talking. Must make noise.

'So...' I pant.

'What?'

'What is the deal with a black bear?'

Charlie stops again, incapable of snowshoeing and having a conversation at the same time.

'The deal is that if it is a black, you fight back; if it's brown, you lay down; and if it's white then you're fucked.'

'Right.' This is his way of saying that you need to play dead if it is a grizzly, and that we don't need to worry about polar bears anyway.

He starts trudging again.

I follow as the sun beats down, turning my puffa jacket into a sauna.

'So,' I call, 'you don't think that there are any bears around here, then?'

'What is this? Twenty questions?'

I can hear the river ahead. Nearly there, I eye the darkness beyond the trees nervously. Should I sing? Charlie might think it a bit odd. Hum? I'm not a hummer.

The trees thin out to nothing and we are at the river. The clear mountain water flows between the deep banks of snow, carrying chunks of glittering ice. It is beautiful. But there is something else here too: tracks, lots of them, weaving in and out, round and about, and they are not beaver tracks. I follow them round. It wasn't just one animal that made these, it was a few at least, travelling together. They look like dog prints, only bigger, each one the size of a grapefruit.

'Wolves!' I say.

Bingo! I'm not worried about wolves – I know they wouldn't attack us, they just don't attack humans – but I am truly on the trail of them now. We might even see one.

'And what is this one here?' I ask, bending to get a closer look at a different type of print, longer with claws on the end.

'That, my dear, is a bear.'

'Oh!' My heart does a funny little leap. So the bears are awake. I scan the tree line again. Suddenly, I'm super-keen again to get the work done and get out of here.

'So what do we need to actually do here, then?'

A brief image flashes before me, our sons in their snow gear, waiting outside school for parents who never arrive.

'Well, I need to work out exactly where I'm setting these camera traps.' Charlie isn't the least bit perturbed – working in dangerous places is his life – and he just starts staking out the river for potential beaver activity. 'Nothing really here,' he mumbles. 'Let's carry on down the river, and we can loop back through the woods and home.'

'OK!'

More trudging. We have to go through the forest – as we turn into the trees, I hum.

'What are you doing?'

'Humming.'

'You're not a hummer.'

'I know, but the bear doesn't know that.'

'There's no bear around here.'

'How do you know?'

'Those tracks were old. Any bear would be long gone, the racket you've been making.'

I'm thrilled that my fieldcraft skills are obviously improving – especially those that relate to survival.

We continue downriver through the trees, as I try to cast all thoughts of predators out of my head. I gaze at the deep blue sky, the pine tree branches trembling under the weight of the pure white snow, yet my palms sweat around my bear spray canisters.

Charlie scales the slippery riverbanks, looking for otter spraint or any sign of beavers, but we find neither.

Suddenly I see an unmistakable movement in the darkness under the trees. I gasp, and my heart leaps wildly. Oh good god, we are going to be eaten!

The large black thing moves again.

'Look,' I say in a slightly higher pitch than usual.

'Oh, yes! Moose!'

'Ah, yes!' I breathe, recognising the mother and her baby.

Close up the moose is enormous. It is hard to make out in the darkness, but I can just see the outline of that strange-shaped head with the droopy mouth. She feels safe enough in the trees to ignore us.

We trudge on in silence for a while. My heart rate slows and my armpits stop tingling.

'You thought that was a bear, didn't you?'

'Of course not.'

But I had secretly hoped for a wolf.

★★★

Days steal by.

We drive for over an hour beneath the Tetons, across the snow-covered National Park, to get to our nearest town so we can stock up on snow boots, warmer coats, a snow shovel and food. There is not a parking meter in sight – bliss.

I get out of the car, zipping up my puffa against the cold, taking it all in, and I smile. Nowhere else is quite like Jackson. Expensive art galleries, tourist t-shirt shops, a shop where you can buy a rodeo buckle, a real-fur G-string for men, a cowboy hat, and the Million Dollar Cowboy Bar with its iconic neon bucking bronco. They all line the central square around which Jackson buzzes; it's a small town with a big town atmosphere.

Nestled at 6,500 feet at the southern end of the valley created by the Teton mountains, Jackson is, at heart, a small cowboy town famous for its dude ranches. Men and women tread the boarded sidewalks in Stetsons, jeans and cowboy boots, not because they like to dress up but because these are their work clothes. Yet modern Jackson has many other faces too: billionaires, cowboys and Patagonia-clad fitness freaks, who all call this place home and rub shoulders with tourists from all over the world who have come here to witness the incredible wilderness, mountains and wildlife. They come to ski or fly-fish, go white water rafting or mountain climbing or just to take in the incredible views.

At each corner of the little, snow-filled town square is a wide arch made of intricately entwined white elk antlers. For most of the year, at any given minute you care to drive by, you'll find visitors taking selfies underneath one of these arches, but today they are abandoned. We have arrived at the end of the ski season, but before the summer season, so the only people here right now are a few locals and us.

Desperate as I am for a good cappuccino, we stop at Cowboy Coffee, even though cowboy coffee is the very kind I hate. Designed, I can only assume, to knock the balls off a concrete elephant – no, sorry, I mean to rouse a cowboy from his sleeping blanket after weeks in the saddle and long nights watching cattle – it's the kind of coffee you can stand your spoon up in. Most Americans drink it by the bucketful. (I'm not exaggerating.) In fact, it is hard to find an American in the morning without the obligatory three-pint insulated coffee mug attached to their hand. To me, it tastes like tar, and I couldn't bear pints of it swilling around my belly. Luckily, due to the influx of billionaires and tourists, coffee has evolved in Jackson and my hopes are high.

The coffee shop is warm. Timber walls are clad in horse art. There is a gathering of cacti on the windowsills and a sign in the bathroom that reads 'Free kitten for every child'.

My husband immediately buys himself a three-pint insulated coffee mug.

While the espresso machine spouts forth – a welcome change from the half-arsed gurgle of the filter machine in our kitchen – I pick up a copy of the free local newspaper. And so begins a new love affair: the *Jackson Hole News*.

On the front page is a story about grizzly bears coming out of hibernation and exactly where they have been seen. I scan it and confirm that Buffalo Valley, our home, is on the list. I have a degree in English and had been thinking about a career in journalism before I ended up talking to puppets on CBBC. In the evenings, I was doing a degree in Ecology, reading with excitement about the reintroduction of wolves in Yellowstone, about conservation plans. The *Jackson Hole News* front page reads like a real-life ecology textbook. I love it.

People around here care about their neighbouring wildlife and the National Park. They follow the elk migration, they want to know when the grizzlies are up, and they know them by name (well, by number). They want to know how high the pronghorn antelope go into the mountains, and just how many wolves are roaming the area. So much of this place revolves around wildlife. I feel myself vibrating with excitement.

I promptly spend too much money at the Jackson bookstore. Books about the history of Grand Teton and of Yellowstone, the world's first National Park; books about local trails; kids' fiction about Native American tribes and children who lived here – settlers, women who turned up in a covered wagon with no warm log cabin or frozen Stouffer's lasagna and somehow found enough food to feed their family through this Wyoming winter. So many stories about this new world.

We all eat buffalo burgers at the Silver Dollar Bar, get supplies and then, having had our fill of civilisation, head home.

Driving through the National Park here is like being on safari in Africa – apart from the snow, of course. It is an American Serengeti. On the way back from Jackson, I spot a dog-like creature trotting across the snow.

'A wolf!' Arthur, in the back of the car, has also seen it.

'No, it's too small, and it has a pointy snout – can you see? It's a coyote,' I tell them.

It has crossed the road and is heading towards the forest. We stop the car. The coyote pauses, silver-grey against the snow, just long enough to look at us and then trots away again.

'You see that, boys? That right there is the way to tell a coyote from a wolf. You see how it moves? If that were a wolf, it would be bigger with a much smoother trot. Coyotes tend to bounce around more.'

As the coyote crosses the stream and disappears into the evergreen forest, I realise I know more about wolves than I thought.

CHAPTER THREE

Ken and Barbie

The valley is coming back to life as the snow slowly melts –
now it is only four feet high instead of six.

Glancing out of the window one day, I see a tanned
cowboy tramping around in the snow outside the small
cabin next door. He is clearing a path. New neighbours!
Excited to catch him before he disappears, I ignore boots,
waterproof trousers and even a coat, and rush outside.

'Hello!' I call.

No response.

'Hello?'

Nothing. Perhaps he is the strong silent type – he is a
cowboy, after all.

I am about to give up when he happens to glance my
way. I wave.

He smiles, revealing teeth to match the snow and his
white Stetson. He is an older man but very handsome.

'I'm Philippa. We have just moved in.'

He smiles again and waves but doesn't reply. I gesticulate,
pointing at the cabin behind me.

'We. Just. Moved. In,' I mouth.

He nods. It's no use trying to have a conversation at this
distance, so I set off towards him and instantly lunge into the
snow face-forward. I had forgotten that the melting snow
will no longer hold my weight, that it is hip-deep, and that
I only have my slippers and a pair of old leggings on.

As I recover myself, I notice his bemused expression. It's
no use, I'm committed now, I can't turn back. So, grinning
and holding his gaze, I struggle through the snow for what
feels like at least half an hour till I finally extend my hand
over the post and rail fence to shake his hand.

'How do you do? I'm Philippa. We just moved in.'

'I know who y'all are. Nice to meet you. I'm Ken and Barbie is my wife – she's inside.'

For the briefest of moments, he glances down at my footwear.

'How are you coping with the weather up here? If we can help you with anything, you be sure to let us know.'

Ken and Barbie. Our new next-door neighbours are Ken and Barbie.

'Thanks. Do you live here?'

'Hmmm?'

'I said THANKS.'

Ken cups his hand behind his ear. 'You'll have to excuse me if I don't always hear ya. I'm a bit deaf.'

We carry on in this manner for a while, with me shouting and shivering and Ken hmmm-ing and pardoning. I tell him about Charlie being a *National Geographic* photographer, and he tells me about Barbie being a horse photographer.

'We'll have horses here when the weather gets better.'

'Oh! How lovely. We have three boys, they will love that.'

I smile. Ken smiles. And then slowly, happy to have chatted to another adult human, I retreat back through the snow.

I am aware that Ken is watching me, almost certainly wondering about this strange English woman wading through soggy, hip-deep snow in a pair of slippers and leggings. Meanwhile, I am trying not to grin too much at the fact that our new neighbours are Ken and Barbie.

CHAPTER FOUR

A lone wolf

Despite their struggle to wrestle free from the grey clouds this morning, the Teton mountains are now gleaming white under a blue sky. I'm getting to know their names: the highest peak, at 13,777 feet and sitting perfectly in the middle of the range, is Grand Teton. The slightly shorter and stubbier one on the end nearest to us, rising over Jackson Lake, is Mount Moran. It was named for the artist Thomas Moran, who was part of the original expedition into the unknown territories of what is now Yellowstone and Grand Teton Park. Moran's work depicting this amazing landscape had a significant impact in Washington. He really showed the magic of the place – so much so that he inspired the desire to protect it and turn Yellowstone into America's first National Park.

This morning the bright, new leaves of the aspens are shimmering discs in the spring sunshine, the grass is deep green, and yellow wildflowers – like small sunflowers – have appeared everywhere. Who can resist such temptations? We are meant to be editing, but something is prodding me and I have to get outside.

'Just a little spin down the lane,' I suggest.

No more than three minutes later, Charlie and I are bumping down the driveway in our borrowed old Jeep Wrangler, under the ranch sign and past the tiny cabins. We slow alongside the community dumpster to see if the local grizzly bear 399 and her cubs are still hanging out there, but there is no sign so we fly straight into the depths of the countryside. It is warm enough to have the roof down, we can hear birds, and as we drop into the forest, we are enveloped in the smell of the pine trees. An old Jeep Wrangler might not seem like a particularly impressive vehicle but – for me, at this moment – it doesn't get much better. Surely we are the closest thing to cool? My blood sings, now in harmony with the overwhelming urge that I had to get outside.

We see him at precisely the same moment. He strides silently from between the straight brown trunks, right into the middle of the road. Silver-grey, huge. So much bigger than either of us could have imagined.

Our first sight of a wild wolf. My dream come true.

I slow the car to a stop, and we are just a matter of yards away. The wolf slows too. Unafraid, taking his time, he breaks his stride and regards us.

We lose sight of him as he saunters into the woods, but as we draw alongside the point where he disappeared, we see his giant silver form blending beautifully with sunlight and tree trunks. He is sitting, waiting for us to approach. I stop the car again.

He seems genuinely interested, examining us with cool grey-blue eyes for long silent moments while we stare at him. His long, pointed face is perfect in its symmetry, his gaze composed and confident. I have the distinct impression that, although he found himself mildly curious about us, he doesn't actually find us very impressive. Having assessed us and decided we are of no consequence, he majestically stalks away, melting into the woods, into his wilderness, leaving us star-struck and gibbering.

We scan the trees, wondering if there are others, but there are none that we can see. Is this my lone wolf? The one I heard? He certainly has the attitude – he belongs to no one. A rare and incredible privilege for a wolf.

I think back to some of the stories I have read from the settlers. Surely there could be nothing more unsettling for them with their fledgling herds of cattle to protect than an encounter like this?

We laugh – after all the work we had put in to find a wolf, we don't even have a single camera with us. But what we do have is a breathtaking image burned into our brains that will last for the rest of our lives.

I am filled with inspiration to find out more and a lingering sense that – despite a borrowed, open-top Jeep Wrangler – we are far from the coolest creatures in the forest today.

CHAPTER FIVE

On the Snake

The wide Snake River is more beautiful than ever. We have canoed down it twice this week, on the hunt for beavers and otters. Fresh snow at the beginning of the week has painted everything white again under the big blue of the sky. We pass the oxbow bend – the halfway point – and still no otters. Not much action of any sort actually. Usually, if we don't see the otters, we see patches of blood on the bank as evidence that they have been eating fresh fish, or signs of spraint. Often we see the grooves down the bank where they have been sledding on their bellies, but today there is no sign in the new snow. We slowly glide down the river, lazy, letting the current do the work as we sip hot black coffee from the Thermos.

'I don't think we're going to have much luck today,' Charlie says from the back.

'Well, you never know,' I reassure him, looking way ahead downriver for any signs of bobbing otter heads.

What I see is better than otters. Adrenaline spikes my system.

'Oh! Oh! A wolf!' I whisper.

Behind me, Charlie grabs his camera, making the canoe wobble. I know this means he has no hand on his paddle.

A large, dark, leggy dog shape strolls down the bank as if he is headed for a drink, but instead he looks back up into the trees behind him.

I bend double to make sure I'm not in the shot, cricking my neck at the same time so that I can still see. Behind me the camera clicks and clicks some more.

The wolf steps into the shallow water beneath him. He looks back up the bank and then across the water again – is he waiting for other wolves to join him? Much as we would like to stop here at a respectful distance, the current is taking us closer to him and – without some seriously distracting back-paddling with both paddles – there is little I can do about it.

'He's going to cross,' Charlie mutters.

'I hope he does it quickly or we'll crash into him. There is nothing I can do,' I murmur. 'I don't have brakes.'

The wolf looks our way – did he hear us? He can't have done, we were so quiet. I realise the wind is at our backs and, even though we are still a good distance away, he must have smelled us. He hesitates as if, for a moment, he is still thinking about crossing and then he turns and trots back up the wide bank. I glimpse a whisper of grey as he disappears back into the trees. Another wolf, I'm sure of it.

We have drifted into the middle of the river, where the current is faster and slipping us down to the wolf crossing-point. We have been so focused on the right-hand side of the bank but now we both realise, at the same time, how many birds are on the left-hand side of the river. Ravens and other corvids are cawing and shrieking in the trees opposite where the wolf was. A bald eagle swoops in low to sit high above them and bends his white head to see what all the fuss is about.

'There's a kill,' Charlie whispers. 'There's got to be – those ravens are fighting over something.'

What we are seeing suddenly falls into context: the wolf didn't want a drink, he was thinking of crossing the river to get to the kill. And he probably wasn't the only one – perhaps there was a pack of wolves behind him. We must have interrupted that. A few minutes later and they would have all been in the water fighting the current like us.

Silently, we float alongside the wooded bank, scanning between every tree. We are both sure that the wolf will come back, skirt around through the trees to get another look at us and try to work out what our intentions are.

A bird starts to chatter in the forest – scolding what? It could be the wolf or it could be us.

'Do you remember exactly where it was?'

'Just beyond the fallen tree.'

We get nearer. No sign of the wolf.

'Pull the canoe in,' says Charlie.

I steer us into the bank and towards the line of fresh wolf prints. As we crunch against the snow, I look into the clear water. On the flat, muddy bottom I can clearly see the large paw prints where the wolf stood as he contemplated whether or not to swim across to the other side. On the opposite bank, I can just make out a dark lump under the trees. Ravens are all over it, magpies flit around the edge of the action and a bald eagle looks on from a high perch. So that is what all the fuss was about – probably a dead moose.

We pull back out into the river. Much as we'd love to see them cross, it isn't our job to keep the wolves from their lunch.

Later, as I reflect on the new wolf image burned into my brain – the dark wolf on the white bank – I remember the final and slightly odd scene in Wes Anderson's *Fantastic Mr Fox*. If you haven't seen it, Mr Fox and his crew encounter a wild wolf that, unlike the rest of the animals in the film, is not wearing clothes and is not standing upright. The scene is incongruent with the movie and the plot is all but done. It seems superfluous and yet, despite being asked, Wes Anderson was adamant that he wouldn't cut it. Most reviewers (and me) seem to think that it is because the wolf represents some unconscious yearning for wildness and instinctive living that Mr Fox has lost by becoming part of society and being responsible for his family. It is even suggested that perhaps Mr Fox was having a midlife crisis and his decision to wave goodbye and wish the wolf good luck is symbolic of him letting go of that 'wild' part of himself. It's no coincidence to me that Wes Anderson used the wolf as a symbol of this.

Having just experienced it, there was mystery about today. As a human, I can't help but come away from the scene with more questions than answers. But I am inspired to find out more. In fact, I am becoming obsessed. I am meant to be here to try my hand at writing fiction. I have lots of ideas but not one of them is quite flowing. Suddenly I know what I am here to write about.

This isn't a new fascination. My second degree was in ecology and conservation. I did it part-time in the evenings, which takes a huge amount of motivation especially when you have been at work all day or when your friends are all out at the weekend and you are the one staying in to write a paper. The only reason I could do it was because I find the natural world – in all its shapes and forms and our place in it and our responsibility to it – so exciting.

One of my particular interests was reintroduction – the idea that we might save a species even when it is lost in the wild. At the same time as I was studying, wolves were being reintroduced to Yellowstone. I was thrilled to see that happen and what this wonderful science could achieve. It seemed to me a symbol of mankind's ability to reverse the damage we have inflicted on the world around us – a symbol of hope. When most of the world was busy ignoring the global warming theory and the accelerated rate at which we are losing species, at least the Americans seemed to have it covered, I presumed. This was one of the reasons I was so excited to come to Wyoming and spend time studying, observing and capturing images of the incredible wildlife in this crucible for conservation.

The first book that I had grabbed to come with us was Douglas Smith and Gary Ferguson's *Decade of the Wolf*. It was as though a part of my brain had been telling me this all along: I would write about wolves.

CHAPTER SIX

Wolves are a problem that won't go away

Our neighbours, Ken and Barbie, have taken us under their wing. It turns out that Barbie is actually Bobbi, and I misheard Ken on account of his rolling American accent, which is a little disappointing. However, it also turns out that they are both great fun to be with. When Bobbi tells her wonderful stories about their lives ranching and looking after her horses, I am captivated. Although her life as a cowgirl must have been tough, it has done her nothing but favours. She is beautiful; her tanned skin speaks of a life lived outside; her wide, lively eyes, her high cheekbones, her smile, energy and appetite for life look like her lifestyle has a much better anti-ageing effect than any over-the-counter beauty products I've been using.

I've always assumed that life on the ranch and range must have been quite a sexist one, with men and women assigned their clear roles, but Bobbi tells me that wasn't always the case. When she was younger and riding out with cattle, she tells me, 'they would send the gals into the huge herds because we were the sensitive ones.'

'What do you mean?' I ask.

'Well, they had worked out that us girls had our special skills. We could see really quickly when a mum or calf was struggling – if one was sick or wasn't bonding in the way it should – far quicker than any man. So we would be sent into the herd to carve those animals out that needed any special treatment.'

After a lifetime of working with them, it is no surprise that Ken and Bobbi are both horse experts, the original cowboy and cowgirl. Now that they are retired, they spend

a lot of their time 'fixing' horses in need. They are true
horse whisperers, which is a gentle and beautiful way to
relate to and train horses.

While I'm happy to chat for hours with Bobbi as she
cooks or tends the fledgling plants on her deck, our eldest
boy Fred spends any spare moment he has with Ken. After a
while, I notice that Ken treats Fred in the same way as he
does his horses – gently but firmly, and with understanding.
Fred thrives in response. Yet a bit of horse whispering
doesn't mean it's all airy-fairy, Ken runs a tight ship.

'There needs to be strict discipline when it comes to
looking after horses,' I hear him telling Fred. 'Every bucket
must always be full of fresh water. Saddles and bridles can't
be left lying on the ground, or they will break. They are
expensive, grooming brushes, and need to be put away where
they belong, or they will be lost. And shit needs shovelling.
Looking after a horse means a lot of shovellin'.'

For our scatty son, inspired by a devotion to horses, these
lessons are far more motivating than any he was learning
at school.

Fred wasn't actually doing so well at school in England.
Only later would we discover that his scattiness was due to a
brain tumour. For now, it is just so good to see him thriving
in the Wild West. Wearing a cowboy hat and boots, his every
spare minute is spent either with Ken, learning about horses,
or with a lasso practising roping techniques on fences.

Ken and Bobbi are particularly good with horses neglected
in the dental department. Ken, it turns out, is also a horse
dentist.

'If a horse's teeth get too long,' he tells me, 'they can't
eat and so, slowly, they starve. Many horses in Wyoming
are left out to pasture or moved in herds to other states for
the winter, and sometimes detailed care like that can fall by
the wayside.'

So it is that one morning, we find ourselves with a skinny-
ribbed chestnut horse tied to the buck-rail fence outside.

'Valentine' can barely eat because his teeth are in such a state. When he does try to eat, most of the food just falls back out of his mouth and ends up on the ground, but we will hopefully be sorting that out for him this morning.

The red-roofed, timber-lined barn is warm and well lit. I breathe in the heady smell of fresh hay. Other than that, the scene is nothing like I have seen before. We lead Valentine to a stall and tie him up. Surrounding him are the kind of tools you might see featured in a Victorian medical manual. I want to pick them up and explore them, but everything is super-clean and disinfected, so there is no touching allowed. Ken is getting himself ready.

'He's going to "float" Valentine's teeth,' Bobbi explains. 'Horses' teeth keep on growing for most of their life. Often they will wear unevenly, and sometimes they'll have sharp edges that cut into their cheeks and gums, making them real sore. Other times it just means that the teeth aren't coming together well enough to grind food.'

A horse's digestive system works at its best when the teeth deliver well-ground food, so this kind of work is essential for good health.

Ken hands Bobbi a syringe, and she gently squeezes it into Valentine's neck.

'Just a little light sedative to keep him calm,' she says quietly. 'Not everyone does this, but we think it is cruel to do this kind of dentistry without.'

We wait a few minutes for the sedative to work, as Ken and Bobbi continue to prepare their tools and, little by little, Valentine's head droops. Although I'm desperate to ask questions, we keep quiet so that Valentine can relax.

Bobbi picks up a rather large metal instrument, a wedge. She bends down to open Valentine's mouth and puts the wedge towards the back. It will keep his mouth wide open and gives us all a great view of his teeth.

Next, her job is keeping Valentine's droopy head up to allow Ken to get to his mouth, so she props her shoulder under his neck. Good job all those years as a cowgirl have kept her fit and flourishing in her old age – I can see how much effort it takes. Valentine is so relaxed that he just settles his heavy head right down onto Bobbi's shoulder.

Meanwhile, Ken has donned a head torch and a pair of magnifying spectacles. He looks particularly odd without his white cowboy hat on – none of us has ever seen him without it before. It feels slightly uncomfortable, somehow unseemly, to see a cowboy without his hat, but we all know better than to comment on it.

He bends forward, squinting and peering into Valentine's open mouth. It is an odd scene, to say the least. After a good inspection, Ken grunts as if he has confirmed his suspicions. Then he starts work, switching on what looks like a Dremel multi-tool, and begins to file.

I see and hear the need for that sedative. It looks brutal, but I guess no worse than a human dentist at work, just on a larger scale. Nothing is said – nothing can be said – as Ken is concentrating hard and the noise of the file is too loud. So we sit and stare at the bizarre scene while Valentine lets it all wash over him, not the slightest bit bothered.

Eventually it's all over. Ken steps back, straightening up with an effort. Careful to put Valentine's large tongue back into position first, Bobbi takes the wedge out of his mouth and gently lets his head droop down.

'Phew!' she says, rubbing her shoulders.

We lead Valentine back out into the sunshine, then pat him and stroke him soothingly until his head is back up and his tongue starts exploring his mouth as if he is wondering what just happened.

Barely a couple of hours later, I look out of the window and see him chomping into a massive pile of food. Happily, none of it lands on the floor. And in just a couple of weeks, he looks like a different horse –no ribs are showing, and he has a gleam to his coat and eye that wasn't there before.

As he gets his health back, Valentine turns out to be a really friendly horse who likes to spend a lot of time with his head over the buck-rail fence, 'chatting' to the boys, who all adore him and race to see him as soon as I get them home from school.

This is another difference in our lives too: I now have possibly the best school run in the world.

In England, I always found the school run incredibly stressful, and when I first saw this one I quaked – no one had warned me about narrow, icy roads, snow five feet deep and precipices. School runs are trying enough without precipices. You are obliged to discuss whose fault it is that the lunch boxes are still on the kitchen table. Or why, after you said 'please go and do your homework,' somebody just 'completely forgot' and ended up playing Lego instead, so that now he has to do it in the car. And usually, my coffee still hasn't really kicked in yet, which means my brain is only just remembering everything that has been forgotten. So I have to multitask, driving at the same time as trying to make a last-minute spelling test fun.

At first, I wasn't sure how I could do all this and then drive in these conditions – I wasn't sure I would have the spare brain space to concentrate on it all. Yet, in just a few weeks of the semester, thanks to the wildlife of Wyoming, my school-run stress scars are healing fast. Together, the kids and I have battled through blizzards to get to school, but somehow, when it's also like a scenic safari adventure, the whole thing is just so much more fun.

We become quite familiar with the moose's iconic silhouette and notice how much of their time they spend in the river, lurking in its bends and eddies. Sometimes they appear on the road in front of us and are in no hurry to move. This year the snow is extraordinarily deep even for them. What a ridiculous oversight in adaptation technology, given that it is snowy here from October through till May.

Can you imagine the internal dialogue of the moose (if it has one) as it is hunted by wolves, stumbling through the deep snow, realising that it has spent millennia evolving entirely the wrong feet for the job?

Most mornings we see a red-tailed hawk sitting in superiority on the timber entrance to the Heart Six Ranch. One day I saw him in exactly the same spot for both the morning and the evening school run. I couldn't decide whether this was some rarely observed ambush hunting technique or just a case of being bone idle. I'd done the shopping, cleaned the house, answered some emails and done two loads of washing while he had just been sitting there pretending to be important. According to my eldest son, who happens to be a falconer, this is just 'what they do'. More like 'what they don't do'.

We were delighted to discover two osprey nests – huge scruffy circles of sticks at the tops of platformed poles. It is rare to see an osprey at all in our normal English lives, and now we knew, come spring, we would see them every day, resident nesting pairs about to hatch a new family. So romantic.

One early spring day, after waiting for weeks, we spot them for the first time, home at last after their winter migration. Yet it isn't quite what we were expecting, and we are slightly shocked to see not two lovebirds but three adult ospreys all in one nest. So much for pairing for life. Eventually one disappears from the picture – I can only hope that affair didn't come to fruition.

As spring progresses, elk begin their annual migration from the elk refuge next to Jackson, where they are fed for the winter, to the higher ground of the mountains. So now our way is often barred by bemused-looking, doe-eyed, giant deer-like creatures with huge white bottoms. They dither – a lot – about whether they should all run across the road together, separately, or maybe just go back. Clearly, elk never evolved a need for snap decision-making – unless perhaps they are being hunted, in which case their only choice is to run as fast as they can. They certainly never seem to have any idea of the urgency of our school run – and I tend to forget our hurry, too, in the moments when I pull to a halt, wind the car window down and watch and listen to and even smell those wild creatures.

Over time, the school run takes on less and less of an urgent feel. We enjoy it too much. Every day brings a new surprise, a bald eagle swooping low over the car, or sometimes our feathery-footed grouse friend. This humble, brown bird about the size of a small chicken has no idea how hilarious it is. For a few days in a row, I have to do an emergency stop when we meet him in the middle of the tight, icy bend just above a cliff, the exact bit of the school run I fear the most. His response to our rather large and noisy car stopping beside him is to freeze as if he has suddenly found himself in an early morning game of musical statues.

'It's because he thinks he is being attacked,' I explain to the boys. 'He thinks that if he doesn't move he is so camouflaged that we won't see him – it's his super-skill.'

The grouse would blend in nicely in the woods, I'll give him that – his tawny shades, huge feathered feet, and brown eyes are not only excellent camouflage but adorable too. But this kind of behavioural adaptation on a snow-white school-run road is really only likely to get you killed sooner or later. Any local still in residence will be driving a lot faster than me.

We don't kill him, of course, but as we slowly edge forwards, we enjoy a little game of chicken with him. After a minute or so of statue impersonation, he can take the stress no longer. Finally, lifting a huge, over-feathered foot, he takes a super-slow-motion step forwards, painfully slowly, then another. When I nudge the car forward another inch, he loudly warbles, 'Run away!' Then, as if someone has wound up his legs, he suddenly sprints across the road in front of us. At least, I'm pretty sure he shouted 'run away'. If it wasn't him, it was one of the kids. Either way, he had us in hysterics.

One day we watch a skunk, trotting down the side of the road just as if it were on a pavement, before it stops and looks both ways then carefully crosses in front of us.

Then there was the best school run of them all. The one where I had to step on my breaks to avoid hitting a family of three grizzly bears who stepped out of the bushes to cross the deserted road.

We know these bears, we have seen them before. The mama is particularly famous – 399, as she is known, is as high as my SUV bonnet. Seeing her so very close, witnessing her size and power right in front of me, nothing would have persuaded me to get out of the car at that moment.

The bears cross the road into a field and break into a run, then a hundred yards away, they stop abruptly to turn and look back at us. For a second, time stands as still as the glistening Teton mountains.

Knowing that we are safe now, we get out of the car. The famous grizzly mama bear and her two yearling cubs observe me and my two cubs in the still, crystal air under the giant arc of the blue Wyoming sky.

'Let's stay. We'll be late for school, but they'll understand,' says Gus, my middle son. He's right – at this school, they will understand, and I want to linger over this moment. After all, we will probably never have a better excuse for being late.

'They are so cute,' whispers Arthur, my youngest.

'They are,' I sigh. From this distance, the family of three looks like cuddly toys, although if we were closer and they wanted to, they could kill us pretty quickly.

'They've grown, haven't they?' Arthur says.

'A bit perhaps.' We only saw them a few days ago in the National Park, so I'm not sure they have really grown that much.

'Is it definitely 399?'

'Yes, it is.'

'Because there are other bears.'

'There are plenty of other bears here.'

In fact, after nearly all the grizzlies in the region were lost to hunters and poachers, now that the bears are protected, Grand Teton National Park and neighbouring Yellowstone National Park have the highest density of grizzlies in the USA. But I definitely recognise this one – she is special. Many of the other bears around here are related to this eighteen-year-old sow, who is an especially great mother. She might not realise it, but she is famous worldwide. People travel for thousands of miles just to see her. After hibernation, her appearance in spring makes the front pages of the local newspaper, as do her cubs and the date she chooses to go to bed for the winter. She often stops traffic in Grand Teton National Park, causing what is known as a bear jam – a traffic jam caused by hundreds of people stopping their cars to happily gawp at a bear. But today there is no one else. Today she has come out of the park to hang out near our house. There are no tourists here, just us. Bliss.

Three honking Canada geese fly noisily between us all.

The bears relax and play a game of rough and tumble, rolling into a pile on the pale yellow-green grass.

'They look like one big bear now,' says Arthur.

The cubs are far from tiny – the family have not long woken up from their second winter together, but they are still not quite fully grown. This will be their last summer with mama. They are kind of on their school run too. Every day is a precious learning opportunity, a chance to experience as many new things as possible, to learn about exploration and survival while mama is still there to protect them before they have to survive alone.

Little do they know how quickly the world around them is changing – there are plans afoot to delist grizzlies, to take them off the endangered species list and to cease the protection they now enjoy, but at least I know that for today no one can hunt this happy family.

399 stands up, shaking off her cubs. Game over. Time to go. I should rouse my own cubs – they also have learning to do – but I can't resist watching her for a few more minutes.

They saunter past a paddock of horses and the cubs are more than interested. Ignoring the wishes of mama, who clearly wants to move on, they squeeze under the fence to investigate these four-legged snorting creatures – perhaps they might be good to eat?

I wonder if we might see a hunt. I'd love to see that, though I'd hate to see a horse killed. The horses have no such intention. They group together and move towards the cubs; far from acting like the prey animals the horses are, I am surprised to see they are on the offence. The lead horses lower their heads, snaking their necks, pawing their feet and snorting ever more loudly.

'Get out of our paddock,' they seem to say.

One of the cubs tries a mock charge.

They charge him back.

Both cubs run. Back to the fence and under it, back to mama who is patiently waiting and watching them learn the lesson of the day: don't mess with the horses.

Mama heads off again, pausing every now and again in the time-honoured manner of a mum waiting for young ones to keep up, then she leads them towards the hills. I wonder if she has a plan for today. She certainly knows where she is going – this Buffalo Valley in Grand Teton National Park, on the edge of Yellowstone, has been part of her patch for over twenty years.

And, for now, it is my patch too – a different place to raise my cubs, to teach them about exploration and survival. How true that phrase is, 'be careful what you wish for.' Only a few months ago, I had wished for a more interesting school run. Now, in no time at all, our lives have changed completely.

'Fancy seeing bears on the way to school, boys,' I say as we climb back into the car. 'What do you think of that?'

'Awesome!' Arthur replies. My son is already turning into the all-American kid.

I am also changing in response to my environment. I am evolving adaptations that are paying off in terms of the survival of my offspring: I have adapted to thoroughly enjoying my school run. Precipices, what precipices?

That evening, as usual, when we get back from school we go straight to see Valentine. Sitting on the buck-rail fence, watching the kids hold up handfuls of hay, and making sure I raise my voice so that Ken doesn't struggle to hear me, I tell Ken about our school-run experience. I know he will be interested in the horses' reaction to the bear cubs.

'I can't wait to see the wolves on the school run next,' I tell him. 'I know they're around, I've heard them.'

'Oh, they're around all right,' says Ken, but immediately I can see he is nowhere near as excited as me. In fact, his face looks significantly darker. He is frowning. 'Wolves, we do not need around here,' he says. 'Everyone is so pleased that they were brought back, and now, of course, they are spreading out of the park. But these people who love the wolves haven't had to live with them, don't remember what it was like living with them. If they did, they might not be so keen. If you'd seen 'em hunt like I have, you wouldn't be so happy either. I've watched them take good cattle down – and calves. I've seen them hunt elk and kill way more than they needed, just leaving the carcasses, just killing for sport. Having wolves around ranches is never going to be a good thing. As far as I'm concerned, they shouldn't be here.'

I'm shocked. Although I know that the reintroduction of wolves was controversial, this is the first time I have actually heard someone I know and like very much – an animal lover – talk negatively about wolves. I kind of assumed that if you loved animals, you would love wolves.

But there was a world before the idea of conservation, and Ken isn't the only one to have a problem with the wolf comeback. Far from it.

CHAPTER SEVEN

Wolves are like Marmite

Much as, I love the conservation and wildlife stories that make the front page of the local paper, I'm also now seeing the problems charismatic wildlife can cause for local people.

Headlines like 'Wolf eats Lucas cow' are pretty common. In this instance, according to the *Jackson Hole Daily*, three or four calves from a herd of 150 had gone missing and then one was discovered half-eaten. A biologist from the US Fish and Wildlife Service, which is in charge of wolf management on behalf of the government, visited the kill site. Based on the tracks around the carcass and the bite marks on it, he verified that it was indeed a wolf kill.

The Gros Ventre pack, who live at the south end of the Grand Teton National Park, are blamed. I read that they have been shot before for taking cattle. In fact, Fish and Game – the government department tasked with the job of managing wolves – nearly killed the whole pack, but by all accounts, one wolf got away. The suspicion is that this wolf has now established a new pack and is perhaps teaching them about the niceties of beefsteak, much to the rancher's cost. The rancher himself has never actually seen a wolf in the 'act of depredation', but there is enough evidence for him to be totally convinced that wolves were and are slaughtering his expensive cows.

Because we are living remotely, our internet connection is sketchy, so after dropping the kids off at school, I head to the town library. As every good writer does, I start the working day with Facebook and Google.

As I see in so many elements of American life – particularly politics – the wolves' social media trail is littered with polarisation. Wolves here are like Marmite is in the UK: you either love 'em or hate 'em. Every comment from someone who loves them is countered by a comment such as, 'The only good wolf is a dead wolf.' For every person who says the reintroduction was a good thing, there are comments from local hunters with theories: Yellowstone reintroduced a non-native species of wolf; because the wolves came from Canada, they are too big for the prey here; the wolves are 'out of control', decimating valuable herds of wild elk and deer and causing problems for ranchers.

The comments and discussions between the wolf lovers and wolf haters are vitriolic. The issues being so passionately discussed are bigger than just the wolves. They spin off into conversations about gun ownership, the president, local versus national politics, trophy hunting, and our relationship with the natural world. The words 'monsters' and 'evil' come up time and again in respect of both wolves and humans depending on which side of the fence the commenter stands. In response to a wolf being shot, I see comments like, 'Smoke a pack or three or four,' and 'One down, the rest to go.' These comments are repeated by more than one person as if they are some kind of anti-wolf mantra or slogan.

So many of these comments depict how people feel about the natural world, our resources and our ability to feed ourselves, and how 'people are much more important than a wolf.' It isn't easy to understand the truth here, but it's clear to see the passion. This isn't just about science; the emotional outcry goes way beyond what I learned in my ecology degree about the science of predator management, understanding a species or conservation. It doesn't seem to matter whether the facts being bandied around are true or false, only that they are being used as weapons in this heated debate. Where people stand on the wolf debate seems to tie into where they generally stand and, at first glance, this argument defines liberals against conservatives, conservation and evolution theory against religion.

Now that I have a better understanding of the current situation online, I scour the Teton County Library shelves. I want answers. I particularly want to know where such deep-seated hatred comes from. There are many wolf books here – probably more than in any other local library – but the most intriguing is a slim, spineless, yet fascinating document. It is a self-published book, all handwritten: *A History of Wolf Times.*

Its author, Doris Platts, was a teacher from the East Coast who moved to Wyoming in 1972 and became a ranch wrangler. I thought I had changed my life quite radically, but that is an extreme life change for a single woman in the seventies. Her picture shows a denim-clad smiling lady, complete with cowboy hat and belt and a twinkle in her eye. I'm sad to see she passed away in 2015, as I'm sure I would have enjoyed meeting her. She was much loved and remembered for one of her favourite phrases: 'Each day is a gift, that is why it is called "the present".'

While living here, Doris self-published ten books or so about Wyoming history, which goes to show the extent of the spell this place casts. *Wolf Times in the Jackson Hole County: A Chronicle* was one of them. On its cover, the large letters of the title sit above a photocopied photograph of a moustached man in classic cowboy attire. Next to him, a dead wolf is strung up on the log cabin wall. Proudly, he holds his gun in one hand and raises the other as high as he can, bringing it just level with the head of the wolf and showing how massive the creature is. What shocks me is the obvious pride and joy he exudes in shooting and 'claiming' these amazing creatures – this is clearly a triumphant moment. It speaks of something so alien to me, a different relationship with wild animals to any that I have known, but it makes me all the more determined to dig deep and try to understand why he would feel like that.

Doris handwrote this 'chronicle' in 1989, six years before the wolves were reintroduced in Yellowstone. Alongside the old photographs, she added her own simple sketches including one of a cowboy lassoing a wolf while on a galloping horse.

'Since there has been so much controversy over the reintroduction of wolves to the ecosystem of our neighbour Yellowstone National Park,' she writes, 'it seemed a fine idea to gather together reports concerning wolves, from the early local newspaper, *The Jackson Hole Courier*.' She includes source material from as early as 1904, and from the *Pinedale Roundup*, which seems to be a source of local news for the ranchers.

Sitting next to the warm fire in the airy library, with a feeling that I have struck wolf gold, I settle down for a long reading session. I'm hoping that dear departed Doris has left me a way to understand the history and context of the wolf-hating here. She ends her foreword by advising me to, 'Draw your own inferences and conclusions from the accounts here gathered.'

The first entry is from 1893. It comes from *Forest and Stream*, the leading American outdoor magazine of the time. Referring to wolves, it says that 'great bands of these gaunt and ever-hungry animals accompanied the buffalo herds.' The wolves were to be found wherever on the continent the great herds of buffalo wandered.

In 1889, in the same magazine, Tomas Seaton observed how quickly wolves learn about traps, poisoning and fire-arms, and how somehow they manage to spread the word to each other and to their young. Wolves are not only clever, I think, but they also protect their families.

By 1905 I can see the hatred for the wolf becoming more apparent – they are described as 'terrorists of the timber' playing 'havoc' with the cattle. A visit to the range is a 'heartrending sight for the rancher as he sees the results of his winter care lying dead'. Cattle are left with their carcasses eaten out and some are 'hamstrung' (when the wolves take a cow down by chewing through its hamstrings, thereby preventing it from running away).

Ranchers take pride in their revenge. Thanks to fresh snow, one rancher was able to track the wolves who attacked and killed three of his cows. He took his dog, a .30 Luger and a Winchester rifle and soon found them. While the dog circled the wolves, the rancher shot and killed each one. I wouldn't believe that one man and a dog could do it, but a photo shows him standing proudly next to all six hanging pelts.

I read about fear, including a fear that the wolves are increasing in number each year, perhaps because the number of cattle is also growing. One reporter writes that they make the nights 'hideous with their howls'. I don't read one good word about wolves. One observer, F. S. Connell, comments that the wolves' 'favorite pastime seems to be biting the calves' tails off short.' Rancher Jess Buchanan is quoted as saying that he has to stay up with his herd all night, but finds that he is in a losing battle – if he is riding one side of the herd, the wolves are attacking the other. The paper suggests that the wolves seem to be getting bolder by the day.

Now that I have spent some time here, I cannot imagine what settlers and ranchers like Jess Buchanan endured to survive. At least, after driving for an hour on a road ploughed free of snow, I can find myself in a huge supermarket and grab a cappuccino. For them, there was nothing, not even the road, not even a car, just a very long winter. I imagine life here with no more heating than a wood fire, the challenge of moving around through the heavy snow, of finding enough food to see a family through the winter, of just keeping warm without Marks & Spencer thermals and Patagonia puffas. I wonder at how difficult it must have

been even to get cattle here, let alone keep them alive.
I know that cowboys brought some herds all the way from
Texas – over 1,000 miles and across four states, battling
severe conditions and Native Americans along the way.

Vast and fertile and no longer occupied by either buffalo
or Native Americans, Wyoming's grasslands were the perfect
place for ranchers to set up, particularly in the summer.
America had an ever-increasing and voracious appetite for
beef, so they had a great market, but no doubt it must have
been tough.

There is no sense in this chronicle, as the Native
Americans had, of living side by side with the wolves. Nor
was there any respect that this is the wolves' land too – and
that was long before the ranchers settled here. There is no
reaching for any understanding, just reaching for the gun.
It seems to be an exhibition of Manifest Destiny – the belief
that America was given by God to the settlers, that this
glorious land was theirs for the plundering, that it was their
right. I guess those settlers never saw it as plundering; they
thought of it as expanding civilisation from the East Coast
through the rest of America, hence missionaries joining the
settlers during the great westward expansion.

'"The great livestock industry", which the government
is pledged to protect, is now suffering from the cruel
depredations of the wolves,' complain the locals. They do,
after all, have to pay grazing rights to the government Forest
Service to allow their cattle to graze there. But they are
clearly expecting 'predator-free' grazing for their money.
They want the government to come up with the cost of
getting rid of wolves. The suggestion is that a man who
owns a ranch and a small bunch of cattle has as much right
to protection from predators as the town citizens have to
protection by the police.

I read tales of 'monster' wolves, seven feet long and
weighing in at over 168 pounds. As the years progress,

I notice the articles get more open in their hatred and in their fear-provoking tone. There isn't any debate about it that I can see; the wolf is the enemy. In the winter of 1915, in the absence of government support, the fired-up locals declare war. One very snowy evening in Pinedale, an hour's drive south of Jackson, the local stockmen meet up. They are so generous in their donations that by the end of the evening there is enough to pay bounties on 'a good many wolves'. The stockmen agree on a sum for each adult of $50 and for each pup $20. However, the hunter will only be given the money if he provides the head and hide of the wolf with all four feet attached.

As spring of that year progresses, more and more reports feature successful wolf hunts. By April, at least thirty wolves have been killed in the area and bounties paid. In October, I am amused to read that Dick Turpin is in town, though presumably not *the* Dick Turpin come all the way from the country lanes of eighteenth-century England to hijack ranchers. Alongside the news of Dick is news of the much-demanded state plans to appoint camps of professional wolf and coyote hunters to live in the woods.

It's around this time that I see the first mention of a 'wolf association'. It is happily reported that the said association is after wolves and coyotes – with 'a vengeance'. There is an increasing sense of triumph over the wolves, as in the following extract dated 1 May 1919: 'Butch Robinson came down from his ranch on the upper Gros Ventre yesterday bringing eleven young wolves which he and his brother dug out of two dens. They shot one of the old ones twice but she got away.'

The rancher's efforts against the wolves are sustained in the name of protecting their cattle. There is not one shred of sympathy for the wolf or sadness at the slaughter. I wonder how terrifying it must have been for the mother wolf as her den was dug out. Did she survive her gunshot wounds?

The likelihood is that she bled to death in agony. The wolves are described as 'profit killers'. It doesn't seem to matter whether they hunt wild elk or deer (which the humans want to hunt) or cattle (which the humans want to sell); there is zero compassion for the 'enemy'.

It isn't just shooting and digging out dens either. By 1923, two full-time trappers are working in the region. That is grim reading to me – trapping seems to be the cruellest way to deal with unwanted animals, whether they are varmints or not. Animals caught in traps are frightened and often left for many hours in horrific pain. Nevertheless, it was undoubtedly a method well used throughout this area at that time – and it is still used now. In those days, professional trappers, known as 'mountain men', spent whole seasons in the wilderness trapping animals for their fur.

The Trapper's Guide was written in 1869 by Sewell Newhouse, a designer and producer of 'the very best' steel traps. His guide was written for 'poor men who are looking out for pleasant work and ways of making money; and especially to the pioneers of settlement and civilisation in all parts of the world.' As well as extolling the virtues of Newhouse's traps, it contains advice on every element of trapper life: the type of equipment useful on a trapping trip, what to eat, what to use as bait, how to skin your animal and then stretch those skins over hoops to get the best price. There are pages and pages of clear instructions on how to trap every animal from skunk to tiger or moose. There is a particularly long section on the intricacies of trapping otters, which breaks my heart as we have had a close relationship with otters over the years and I know them to be fun-loving, family-loving and incredible hunters perfectly adapted to their habitat, and I have enormous affection for them. And, of course, the wolf is included in this trapper's guide.

I'm struggling to understand the trapper's attitude that all of nature is there for his profit – even the joyful otter – and yet I need to if I really want to know why there is such a polarisation of views about the wolf. Despite trying to be balanced, everything in me is repulsed as I read, 'when roaming singly, they are sneaking and cowardly.' What about ingenious survivors? 'They are carnivorous and combine both ferocity and cowardice in their character.' That makes me fume – how much more cowardly is it to lay a trap? I realise that I'm doing the very thing I am interested in avoiding: I'm being one-sided, polarised in favour of the wolf.

Newhouse writes, 'Troops of them [wolves] have been known to pursue and attack men.' Although in modern times, wolves do not generally attack humans in North America, I recognise that, at that point in history, wolves hadn't yet learned to fear men and guns. So how do you trap a wolf?

The wolf trap apparently gives off a scent of the iron from which it is made. To disguise that, the trapper is advised to cover it with blood. 'Which can be done by holding it under the neck of some bleeding animal.' Alternatively, he could smoke it over hemlock or cedar boughs. The 'clog' of the trap, the weight, should be at least fifteen or twenty pounds and the whole thing should be disguised under soft earth or ash, and any trace of footprints removed with a soft quill or bush. Then, 'to make the allurement double sure, obtain from the female ... the matrix in the season of coition and preserve it in a part of alcohol tightly corked.' What is this matrix, and how does one get hold of it? I can only imagine, and it probably isn't pleasant. The guide advises the trapper to scent the bottom of his boots with the matrix to make a trail leading to the trap. Trapping turns out to be an intricate process demanding quite a bit of skill and labour: 'Care should always be taken to keep at a proper distance when looking after the trap, as the wolf's sense of smell is very acute and enables him to detect the footprints of the hunter with great sagacity.'

The trapper's life was not an easy one. Way out in the wilderness, they endured all weathers and conditions and were alone a lot of the time. Lines of traps might be as long as forty miles, and the trapper must walk the line to check them. I know from my own forays into the Wyoming wilderness that this would be no simple task. Conditions are continually changing here but always extreme. If the weather has been a little colder yet the snow is deep, for example, I can find myself 'potholing' with each step – the weight of my body and the kit I carry easily crushes through the surface layer of ice and I end up in snow up to my thighs, making every step from then a considerable effort. In the summer it isn't an easy stroll either – temperatures can get really high, and at this altitude the sun is fierce. There are streams to forge and hills to climb, and making your way through the sagebrush, which covers the ground like broom and heather, makes it all the more difficult. Sometimes wildlife trails meander through the sagebrush. This can be helpful if they happen to be going in the same direction as you, but then, of course, there is the wildlife itself to consider – bears, moose and bison are all dangerous, especially if you take them by surprise.

Osborne Russell ran away from home in Maine at sixteen and became one of the most famous 'mountain men' in this area. He wrote the *Journal of a Trapper*, which – once you get your head around the killing – makes for extraordinary reading. This was no coward. He describes vast rendezvous of the 'whites' and 'Indians', filling whole valleys, to trade and sell furs. At one of these, the first two white ladies to 'penetrate' the rocky mountains arrived, on their way to establish a mission with the Native Americans. They were 'gazed on with wonder and astonishment by the rude savages.' By page twelve, he tells of an encounter with a grizzly bear that nearly ended his life when it attacked him with its 'enormous jaws extended and eyes flashing fire'.

He describes Jackson's Hole, as it was then known, as rich in game and fish. When his trapping party heads north, crossing the Yellowstone river their horses are nearly swimming. They hunt in what is now Yellowstone National Park – nowadays filled with tourists, boardwalks and signage but in those days a wilderness, not yet settled, and the geysers and hot pools for which Yellowstone is now famous must have been a bit of a shock. Osborne writes how difficult it was to get around 'whilst the hot water and steam were spouting and hissing around us in all directions … Shortly after leaving this resemblance of the infernal regions, we killed a fat elk and camped at sunset.' That night, wrapped in blankets against the deepening winter, he listens to the howling of a solitary wolf who 'bewails his calamities in piteous moans.'

Osborne's journal shows, with no romanticising, just how tough a trapper's life was; this was not a life choice for a person without true grit or, as Newhouse says, someone looking for 'pleasant work'. He survives gory encounters with Blackfeet Indians, when the earth seemed 'teeming with naked savages', and makes good friends with other Native American tribes. Osborne describes how the trappers made themselves 'lodges' from buffalo skins in which six of them could huddle together to survive the cold. After travelling twenty frozen miles in just one day to trap beaver, for example, he and a colleague found so much snow and ice that it was impossible to continue, so they ended up staying in a large cave for six days and 'making havoc' with the buffalo instead. Another night, the snow and wind were coming so fast that there was no place to bed down other than by digging into a cave that had been washed out by floodwater in the spring. He and his friend managed to make a small fire of sagebrush, and their dinner was a small piece of mutton in one hand and a piece of snow in the other. He then 'laid down to shake, tremble and suffer with the cold till daylight.' Doesn't sound like pleasant or easy work to me, but presumably, it must have been worth it.

Osborne's descriptions of the land through which he travels, and in which I now live, are beautiful. It's clear it touched him as it touches me, and I find his stories of the different Native American tribes fascinating, but there is no doubt of the brutality of the conditions here.

I don't see the same hatred for the wolves as I see in Doris Platts' collection of articles. Yet there is clearly no emotional attachment to the game and wildlife he kills either, and there is no sense that nature's bounty would ever run low. His work is part of a large-scale business with large teams of men involved and big companies. It is only after nine years of trapping and hunting, and an unusually severe winter, that he comments, 'trappers often remarked to each other as they rode over these lonely plains that it was time for the white man to leave the mountains, as beaver and game had disappeared.'

Shortly afterwards, Osborne himself resolved to leave and adopt a safer, perhaps less exhausting, life. On 22 June 1842, for the last time, he gazed upon this place he had known and loved and where he had experienced so much. He was moved to write 'The Hunter's Farewell', a poem in which he says goodbye to what – even in those days – was an extraordinary lifestyle. This is just a small extract:

> The prize obtained, with slow and heavy step
> Pac'd down the steep and narrow winding path,
> To some smooth vale where crystal streamlets met,
> And skillful hands prepared a rich repast;
> Then hunters' jokes and merry humour'd sport
> Beguiled the time, enlivened every face.
> The hours flew fast and seemed like moments, short,
> 'Til twinkling planets told of midnight's pace.

But now those scenes of cheerful mirth are done,
The antlered herds are dwindling very fast,
The numerous trails so deep by bison worn,
Now teem with weeds or overgrown with grass;
A few gaunt wolves now scattered o'er the place
Where herds, since time unknown to man, have fed.
With lonely howls and sluggish, onward pace,
Tell their sad fate and where their bones are laid.

These last lines speak of the change of fortune for the wolf at the hands of the 'whites'. Osborne has observed the decimation of the buffalo and elk herds, and the consequences for humans and animals alike; for the Native American and the wolf, life here would never be the same.

Osborne goes on to work with settlers constructing sawmills and, after an accident in which his right eye is blown to smithereens, his journal comes to an end. But he has told me all I need to know. Between the years of 1834 and 1843, one man's journal tells us that the region's game went from being so plentiful that it was perceived as God's gift to man, to almost disappearing.

Osborne wasn't your average 'mountain man'. He was described by a friend as a man with 'refined feelings', which I guess was the inspiration for his journal. I'm glad he was, but ultimately it didn't make much difference to the wildlife. It wasn't until 1894 that Congress took any notice of the buffalo's plight. In the 1830s, before buffalo hunting truly began in earnest, they roamed in their millions; by the time Congress took action and protected them in the National Parks, only three hundred were left.

And so where could the wolves go from here? If the herds of buffalo were still roaming, would the wolves have any need to attack the cattle that replaced them?

In Of Wolves and Men, Barry Lopez argues that there was more to wolf killing than just protecting cattle. His theory goes that the wolf hatred has religious and secular roots and that wolf killing is somehow 'righteous'; that perhaps the wolf represented the wilderness and all that the settlers and ranchers – with the idea of Manifest Destiny – were trying to conquer in the name of civilisation. The wolf stood in the way of progress. Besides, throughout our history, in myth and fable, the wolf has largely been demonised. Lopez says, 'It was against a backdrop of ... broad strokes – taming wilderness, the law of vengeance, protection of property, an inalienable right to decide the fate of all animals without incurring moral responsibility, and the strongly American conation of man as the protector of defenceless creatures – that the wolf became the enemy.'

According to *Wolf Times*, Ben Goe, who lived on a ranch in the elk refuge just north of Jackson town, lost a colt to a wolf attack. He describes the poor thing trying to get away by using only its two front legs as the wolves tore at its haunches. Another horse of his took two years to recover from a wolf attack because they had ripped its haunches down to the bone. Ben reported hearing his last wolf in 1916. I can understand why he was relieved.

The US Fish and Wildlife Service officially stated that the last wolf in Wyoming outside Yellowstone was killed in 1927. Many years later there were two more sightings south of town reported at the same time, but they were probably one wolf just passing through.

The large-scale killing of wolves in Wyoming was repli-cated in other Rocky Mountain states including neigh-bouring Montana. The upshot of this sad slaughter was that in 1973 wolves were formally listed as an endangered species. It is clear to me that what I'm reading about wasn't predator control, it was eradication. There was no mercy for the wolf. There was no anticipation of the ecological consequences of the decimation, although at that time, there was no understanding or appreciation of ecology. And from Osborne's journal, I see there was not really a sense of finite resources either.

I look at the photo of Doris Platts one more time, her words from the foreword echoing in my mind: 'Draw your own inferences and conclusions.'

What must it have been like? To bring cattle and cope with the extremes of weather with only basic provisions and accommodation. To believe that, in God's eyes, you were doing the right thing by settling here and transforming wilderness to civilisation. To have God as a source of unquestionable conviction in your behaviour, which, in turn, gave you courage and strength. What must it have been like to hear mysterious wolves howling at night and fear them? Fear that you might be powerless to protect the stock you worked so hard for, the animals you needed to survive. How easily that fear turned to a hatred of the swift beast of the forest.

As a race, we almost certainly were not doing the right thing. When I imagine the cruelty and suffering inflicted on the wolves, I am sad beyond belief. Yet – in the context of those individuals who were trying to do what they thought was the 'right' thing and protect their hard-won livelihood – when I really push myself to understand the opposite point of view to my own, I get glimpses of why they saw persecution and eradication as the only option and so I have a much better idea of where the present-day 'wolf hatred' originates.

Only with empathy and by understanding the other point of view will we ever have the power to change things, to save things, including our world. If studying conservation has taught me anything, it is that.

So what is the world of the wolf? What do we know now that we didn't know then? Does it really matter whether they exist or not? And why do I care so much?

CHAPTER EIGHT

Wrolf

The text message simply reads: *Wolves headed west from Mormon Row. 7–8 black.* I read it on the phone shoved around the bathroom door as I get dressed and ready for a lunch date with my husband. So I ignore the nice cashmere sweater I had planned to wear, and scrabble for all my warmest layers – none of them matching but who cares? With no time to dry it, I scrape my wet hair under a hat. Into the car we bundle camera kit and cross-country skis – just in case we need to trek. I cancel a doctor's appointment and pray we will be back in time to pick up the kids.

Please let this be it – the day we have a good wolf day. The day we finally get those elusive shots of wolves.

We head for Mormon Row, a particularly photogenic part of the National Park, where a group of Mormons arrived in the 1890s and settled in a community of homesteads. According to the Homestead Act of 1862, settlers were entitled to up to 160 acres of Wyoming land on condition that they live on it for at least five years and 'prove' it, which meant improve the land. John Moulton and his brother arrived at this spot in 1907, settled and cleared over eighty acres of land. They dug wells for the community and built corrals, a barn and cabins. Later, John built a house for his family, which he painted pink for his wife while she was in hospital. She never really liked the colour, but she was so thrilled with the idea of the 'gift' that was meant to delight her that they never changed it and it is still here today.

Romance aside, my favourite building here is the much-photographed John Moulton barn – a sleepy, dark-timber building in the traditional shape of a central section with a smaller portion at either side and the whole blanketed in gently sloping roofs. These roofs are pleasing to the eye, but they were mainly designed to cope with the massive snowfall in winter. The barn stands humbly against the backdrop of the Tetons, a constant reminder of their elemental setting. In the deep snow today there is no sign of the irrigation channels he helped dig and which enabled the locals to thrive through the hot summer after drift on drift of white winter. The barn took Moulton over thirty years to build, although he did live to the ripe old age of 103, so there must have been some benefits to such a gruelling and physical lifestyle.

We drive to where the snowplough makes an abrupt seasonal end to the long road. This road goes for miles into the sagebrush scrub in the summer, and it is fun to drive out and spend time with the large bison herd that hangs out here. Now the only clear roads are those that lead to houses. Even the mountain tops are craggy today. They frown at us from behind grey shrouds and then cover themselves again.

It isn't easy getting out of the warm car. I struggle to open the door against the snow-slapping wind that then buffets me as I scan the white plains with binoculars, willing every smudge on the landscape to move, to stand, to run.

'What are y'all looking for?'

A brown-faced, white-bearded man, his ears covered by an enormous black furry hat, appears as if by magic, like the shopkeeper in *Mr Benn*. We might be asking him the same question – why does he walk through the harsh wind just to get to the massive pile of snow that now punctuates the end of this road?

'We're just looking to see what there is,' I say. 'How about you?'

'Just walking. Need to get out whatever the weather,' the man smiles, then turns around to walk back to wherever he came from.

There is nothing to be seen. We turn back down the road and have a conversation about whether the wolves might have switched direction. It is unlikely they would attempt crossing the main highway in daylight and head west – but if they did we would never find them – so perhaps they followed the riverbed east into the wilderness. Our only option really is to drive around the back of Blacktail Butte and head south to Kelly, a small 'village' nearby. As we get to the cluster of wooden buildings set in the antelope flats, I am reminded of the stories I read about the people who lived here, of Butch Robinson who tracked the wolves up into the hills with only his dog and his gun. Like him, we are tracking wolves today but with a camera.

Charlie spots the wolf. A black dot in the distance on a ridge of snow. As we approach, we see another car stopped beside the road. A man has also seen the distant wolf and is walking in a direct line towards him. He has no camouflage clothing, a tiny camera and clearly no fieldcraft but he is still too far away for the wolf to have noticed him.

We sit quietly, taking the time to watch the wolf through our binoculars, wondering if there are others. He appears to be sleeping alone, curled up nose to tail, high on a ridge of snow. As the vicious wind rocks our car, I can't help thinking that it doesn't look like a great place for a nap, but he is clearly very relaxed. Where are the others? The message mentioned at least six. They are probably still nearby.

The man disappears into a gully.

Charlie gets out of the car, quickly dresses in white and grey snow camouflage, suddenly looking like something out of *Star Wars*, and grabs his camera. I stay inside the vehicle, as two of us walking towards the wolf in the snow are far more likely to disturb the wolf. And I get the bonus of keeping the heated seat on while I watch my husband make his way across the snow.

The wind buffets, my husband creeps, time passes. I wonder if it is the howling of wind or wolves I can hear.

The question remains on my mind: why do I care so much about wolves? When I contemplate it, I think I can trace this infatuation back to childhood. Not only did I have a fascination and passion for animals, but also for fiction. I would escape into it. Sometimes – well, often – the world of fiction was better than the real world. My favourite book was *The Little White Horse* by Elizabeth Goudge. I read it multiple times and at various ages – it was a form of comfort. One of the things I loved about the central character, Maria, is that she is friends with many animals. These aren't talking animals, but real ones that she develops an understanding and friendship with, and these 'friendships' support her as she develops and resolves the conflicts in the adult human relationships around her. One of these animals is Wrolf, a huge 'dog' who is actually a lion. As I child I didn't like that image, and it didn't really go with the name, so for me, Wrolf was always a wolf (although, at that point in my life, I had never seen a wolf except in my imagination). He is independent and mysterious but loyal, caring and protective. Something about his size and strength reflects Maria's strength of spirit but balances her physical frailty, and something about his guardianship appealed to me. As a child, I longed for a Wrolf of my own – with a Wrolf at your side, you would always be OK and protected. So I guess, although Wrolf was really a lion, from early in my life, it's clear there was a link between wolves and benevolence, protection and friendship in my imagination. This was only reinforced by my understanding of dog domesticity and the fact that humans and canids have always coexisted well and had deeply bonded relationships.

Only later, as I was becoming more and more interested in ecology and conservation, did I come across a quote by Goudge, who, unlike me, was a Christian: 'Nothing living should ever be treated with contempt. Whatever it is that lives, a man, a tree, or a bird, should be touched gently because the time is short. Civilisation is a neater word for respect for life.' Was it through reading *The Little White Horse* that my beliefs aligned with hers, or were my beliefs the reason I loved the novel?

We humans have always had a relationship with wolves through literature. The wolf is a symbolic figure – usually a negative one, but not always. Our myths and fairy tales are full of them. Romulus and Remus, the founders of the Roman Empire, were said to have suckled at the teats of a wolf. Kipling's Mowgli has a wolf, Raksha, as his adopted mother and grows up as part of the wolf pack. Little Red Riding Hood's adversary is a wolf in a cloak. And of course, we all know the one about the wolf in sheep's clothing and the one that blew the little pigs' houses down.

As a scientist, my positivity about wolves has increased now they are a conservation success story – who wouldn't want the wonderful wolves back where they belong? But I wonder if I am already approaching the subject through the story I have told myself, one that has built up in my imagination. I make the decision to find out. After all, until now, I have never actually lived with wolves.

Suddenly the giant dog is up, taking steps, soot-black on the snow.

I can barely see Charlie; he is low down and camouflaged. But something has spooked the wolf. He canters along the ridge, glancing behind. Perhaps it is the other wolves. He stops and turns then walks back – slowly, deliberately. What can he see? He turns again and starts to run.

I'm sure I can hear barking in the distance.

Charlie appears again, making his way back to the car, his feet plunging into deep snow. My heart sinks – he is giving up.

'Did you see that idiot?'

What I couldn't see from the car was that the visitor had carried on walking in a direct line towards the wolf, who couldn't hear him for the wind and had only caught a scent of him when he was really close. The wolf had been taken by surprise and been terrified.

One surprising part of fieldcraft I have learned is that it isn't always best to sneak up on an animal. Sometimes it's best to let it know that you are there, but that you are not a threat, that it can relax in your presence. Part of living in a National Park is realising that not everyone has the same understanding of animals that we do – for example, the people who try to put their kids on the bison for a photograph, having no clue of the danger they are in. Or the people who every year ask 'When do the bears get put to bed?' For some people the idea of wildlife is not real – they see the National Park as just an extended zoo.

That old idea that wildlife is there for human enjoyment and entertainment still thrives.

CHAPTER NINE

Fight or flight

Over the next few weeks, I'm sad to discover that traps are still being used in Wyoming – and even sadder to learn that sometimes dogs are caught up in them. In 2018, a beloved pet Pyrenees on her regular walk with her owner was enticed by the meat in a bobcat snare that killed her just twenty feet from the trail used by dog walkers all the time. Trapping is still legal in this state.

In 2012, Lisa Robertson founded Wyoming Untrapped an organisation that campaigns against the 'archaic' trapping laws in Wyoming. I first came across them when I picked up a leaflet about how to extract your dog from a snare or trap. I was amazed to read that trapping is legal on 85 percent of public land. Traps are indiscriminate: one trap set for wolves caught a grizzly bear, which is still on the endangered list. She was sedated and released from the trap, but the degree of damage to the trees and vegetation surrounding the trap showed the amount of distress she was in while she 'waited' for that release. In 2015 a wolf trap was discovered in Montana with a severed mountain lion paw in it. Claw marks ravaged the trees around it, and it is supposed that the lion ate off his own paw. It is unimaginable to think of the suffering that motivates an animal to do that to itself. Although it may have survived for a while afterwards, there is no way a big cat could hunt in that condition, and it almost certainly will have died. Mountain lions are not protected, but there is a 'quota' for them, which means that any hunting is limited. There is no requirement for mountain lions trapped by mistake to be reported, so their deaths are not added to the quota.

When Lisa and I meet, I adore her immediately. Five of us are crammed in a car, all buzzing with excitement. We initially met in response to a crisis: grizzly bears had just been delisted, and Wyoming state officials were putting a hunting season in place for them. This included a zone around the park covering the area in Buffalo Valley – an important part of 399's territory. We were aghast. It would mean that as soon as she set foot over the park boundary with her cubs, she could be shot by a waiting hunter. She is so famous that not only does she draw tourists from all over the world, but also people know where she can be found most of the time.

Needless to say, many people were unhappy with the decision to hunt grizzlies – not just for 399's sake, but for all the grizzly bears in the area. So Lisa and her friend Ann – both women who put action behind their words – created a plan. To be able to hunt a bear, you would have to apply for and be issued with one of a limited number of tags issued by Wyoming Game and Fish Department. Lisa and Anne got to work to raise money so that anyone who believed grizzlies shouldn't be hunted for trophies could be funded to apply for a tag and go and shoot the bears – but with a camera, not a gun. The campaign went swimmingly and one of my neighbours, Kelly, successfully applied for a tag, which meant one hunter down!

During our conversations, Ann and Lisa quickly realised that my passion for wolves matched theirs and so they offered to take me to a wolf den. And so here we are, early one warm spring evening: a car full of post-menopausal women filling each other in on the latest gossip, albeit about wolves and grizzlies rather than people.

The journey involves a fair amount of off-roading, including one 'Woah, that was deeper than I thought' moment as Ann very capably drives us through a river. Then we have a steep hike, during which my face, ears, and even eyelashes are bombarded constantly by tiny black flies. I learn very quickly not to breathe with my mouth open – though that doesn't stop them from going up my nostrils.

Eventually, we reach a broad grassy ledge overlooking a lake, from which you can see for miles. It strikes me that, although I have climbed most of the ridges I can see with a heavy backpack full of Charlie's camera traps, we have never explored this area. We set up our scopes and applied bug spray, which doesn't give us any relief at all – probably because by now we are all too sweaty – but the scopes work, and that is more important.

I know this has been a regular den site for years. In fact, it was the pack's original den when they moved into the area. None of us knows whether they will be using it this year or not, but the first thing to do is to locate it. That sounds easy, but it isn't.

'So,' says Lisa, 'I remember it was below that triangle of trees.'

Looking ahead, I see a perfect loveheart of trees, so I focus my scope on that and come down a bit.

'Then left a little and it is right there in the gully,' Lisa continues.

I can't see the gully, of course, because it is covered in shrubs and because it is a gully.

'That is the original den, but there is another further to the right – a new one. I don't know which one, if any, they might use this year.'

'I don't see anything,' I say.

'Me neither.' Lisa's voice is soft with the twang of a Southern belle. 'We will probably have to wait a bit. They may be there but not ready to go hunting yet; it is still quite early.'

The sun is moving slowly down to the horizon, and I wonder if the wolves might wait till dark to emerge. I can't quite imagine doing that hike in the dark.

'I hope we see them tonight,' says Lisa. 'I worry about them. I wonder if they were shot this winter for the cattle depredations. I know some were.'

Although she never stops working for what she believes in, there is a sadness in Lisa, a sense that she has done all she can do but that it will never be enough.

'Hope hurts,' she says. 'When are we going to *get* it?'

'What do you mean?'

'I worry that we aren't progressing in the right direction. Delisting these animals, making them available to trophy hunters, is like going backwards. I truly hoped we had moved on from all that.'

Lisa's slight build belies the strength of her personality – she is older than me but probably has three times the energy. Before Wyoming Untrapped, her passion was for wolves in particular. It still is. Originally from North Carolina, she has lived in Wyoming since 1991, when she and her husband built a house that they still live in today. For much of the time since then, she volunteered with the wolf recovery programme, but she admits she is no biologist or scientist.

'What started it all off then?' I ask since we seem to have a while to wait for the wolves.

'Well,' she begins, her voice gentle and rolling, almost hypnotic, 'My husband and I were actually working with a wildlife expedition company. From there, we got to know more and more biologists, and one day I was chatting with one of them, a guy called Tom Segerstrom. That day he was clearly frustrated, so I asked him what was wrong and he said, "I just can't find my porcupines."' Lisa giggles. 'You see, he was trying to do a research project on porcupines and he had two of them collared with radio collars so that he could track them. He had just come back from many hours traipsing around in the wilderness with no luck at all. So I had an idea. You see, I'm a pilot.'

'You're a pilot?' I'm surprised. Lisa seems like the least 'piloty' person I can imagine.

'Yes, I have my own plane. I learned to fly in my late twenties,' Lisa says, ignoring my surprise, as I start to realise that her story is never about her – she has far more important things to talk about. 'So that's where it started, really. In those days, I had a Tiger American General. It was a single-engine plane with a slide-back roof, quite fast. So I took Tom up in it with his antenna, and we had those porcupines tracked in ten minutes. I had saved him possibly days of tracking on the ground. That's when I realised that I wanted to find things I could do with the plane that would be helpful to wildlife and research, and that is really where it started. Over the next few years, I got involved with all sorts of projects and students. It was such fun – we tracked everything and anything: elk, lynx, grizzlies, even trout!' She laughs at the expression on my face. 'Yes, we even tracked Clark's nutcrackers.'

'Seriously?'

'Yes.'

'From a plane?'

'Absolutely!'

We both laugh. The Clark's nutcracker is a small bird, a little smaller than a crow but part of the same family. Dove-grey with black wings that have a flash of white, it has a black eye and black beak and really is lovely. It is significant conservation-wise because it is the only bird that spreads the seeds of the whitebark pine. Whitebark pine forests cover the Rocky Mountains, and their future is under threat from the warmer temperatures caused by climate change. The mountain pine beetle and blister rust also kill them. Mature, cone-producing trees provide a high-quality food for wildlife, especially bears and squirrels as it is 52 percent fat. Three quarters of whitebark pine trees died between 2002 and 2012. Studying the bears as they do in Yellowstone, the biologists have already observed a change in behaviour: bears are spending on average a week less in the whitebark pine forests since the decline began, and they are already switching to a different food source. Studies seem to show that male grizzlies are eating more meat – even if that means scavenging wolf carcasses – while females are eating more truffles.

I watch a group of pronghorn trotting away from the area up the opposite hill.

'Do you think that might be a sign? Might they know the wolves are there?'

'Could be. That was another of the things we did with the plane actually – we followed pronghorn and elk and discovered their migration corridors.'

That research, which Lisa so casually drops into the conversation, was instrumental groundwork; discovering the migration routes for pronghorn led to the protection of these areas. Pronghorn are not antelope, but they look just like them: small and sand-yellow to light brown with horns. Whenever we drive around Wyoming, we see groups of them grazing by the roadside. They are built for speed like the antelope we see in Africa. They are so fast that they can hit 60mph from a standing start and cruise at 45mph, which gives them the well-earned title of North America's fastest mammal. I assumed they would be a valuable prey species for wolves, but the truth is that coyotes hunt them more. The wolves actually do the pronghorns a favour because, according to recent studies, the presence of wolves keeps coyote numbers down, allowing pronghorn numbers to go up. The rule of the wolves in our ecosystem is more complicated than we know, and we certainly wouldn't know as much as we do now if they hadn't been reintroduced into the view of the scientific community of biologists working in and around Yellowstone.

The more I discover about Lisa, the more I am intrigued.

'After a while I decided to trade my Tiger in for a Cessna 182, which was safer for mountain flying, more stable,' she tells me. 'And while I had the chance I got the FAA to allow it to have a restricted category so that I could permanently have wildlife antennae in the back of the plane. No one had done this before, it was all a self-taught learning curve, but pretty quickly I could rig in minutes and have all the receptors tuned to all the frequencies of the different animals. It was a high point of my life,' she smiles. 'I lived and breathed it. I was flying every day of my life and I took on any project, including flying people down to monitor the progress of the new oil rigs in Pinedale and for them to see what that was doing to the sage grouse habitat.

'At that time the wolves were just being reintroduced up in Yellowstone and I was dying to "fly" with wolves – well, you know what I mean, I call it flying with them. A couple of years after their reintroduction, the wolves were starting to filter south of Yellowstone so I got in touch with Ed Bangs. He was the wolf recovery coordinator for US Fish and Wildlife at the time. I asked, "how can we help you with your work with our plane?" So we talked to Doug Smith, head of the wolf project up in Yellowstone, and asked him what he needed – because once those reintroduced wolves left Yellowstone they were no longer under Doug's authority. We worked out a plan where I could research wolves and report to Ed and Doug. There were a lot of legalities and a lot of paperwork, but I was finally wolf tracking from the air. I could report on where they were, and we could also report to ranchers and let them know where the wolves were.

'It was an amazing time for me. I did it for ten years – I know the families and the wolves that were born here. I still have the radio collar of 24F, the first female that made it down here. She was amazing and I still know her frequency off by heart.' Lisa has a wistful look in her eyes as she recites it. '217.000. It was part of history. It makes me emotional now. I'd see them from the air. She was this beautiful black wolf and I swear she knew the sound of my plane.

She'd glance up at me and then just carry on with what she was doing. I just loved it.'

That wolf, 24F, was one of the first to venture out of Yellowstone National Park and into Grand Teton, and she was pretty famous for it. She came from the Soda Butte pack, one of the original packs formed when the wolves were brought from Canada to Yellowstone and released into temporary enclosures of about an acre each. The Soda Butte pack were christened that because their new home was right at Soda Butte Creek near Silver Gate on the east side of the park. Biologists selected an alpha male, an alpha female known as 14F, and two others to form the pack, then the four of them were placed in the enclosure together in mid-January and fed roadkill and carrion while they worked out their hierarchy and acclimatised.

When the Soda Butte pack were finally free to enjoy the freedom and protection of the world's first National Park, they promptly decided Yellowstone wasn't for them and left, setting up home in the Beartooth Mountains just north of Cooke City and Silver Gate. The alpha female, 14F, got pregnant for the first time and that summer she found a den in a remote place, but sadly only one of her pups survived: 24F.

The pack dispersed. Two males went to Wyoming and were shot. One, 12M, made it as far as Daniel south-east of Jackson, which — if he took a direct cross-country route — would have been a journey of around two hundred miles. The following year the pack set up home on a 13,000-acre private ranch where they killed nothing but elk. They didn't bother cattle once, but the human neighbours were not impressed, and death threats were made. The biologists needed to protect the wolves and, although they were not happy about it, they decided to recapture and once again relocate them within the safety of Yellowstone Park.

At this time there were four adults and four pups — including the yearling 24F. They were placed in another pen for the summer of 1996 in Crystal Creek in the south-east of the park, where there were no other wolves. Not long after they were released the alpha male, Old Blue, died of old age. Number 14 took off on her own for a while far into unknown territory before returning to her pack and, while the biologists couldn't be sure, it seemed as though she might have been through a period of mourning.

That winter, the bad weather meant a scarcity of elk, and so the wolves were forced to move into the territory of the Thorofare pack. They were not welcome but were triumphant in the ensuing battle, giving them a massive territory that ranged south past Yellowstone's southern boundary. Although it was huge, however, this was not a food-rich territory. By 1997 they had started to explore by following the elk, and 24F was seen in our stomping ground, Buffalo Valley, before heading back into Yellowstone. The following year they were bolder still: in 1998, just three years after reintroduction, the Soda Butte pack had followed the elk herd all the way to the National Elk Refuge just outside Jackson.

On their way back to Yellowstone, 24F got together with a yearling from another pack, 133M, and stayed with him south of the park. Together, they were the first wild mating pair of wolves in Grand Teton National Park, so they were dubbed the Teton Duo and established themselves right near where our kids would later go to school, in Moran on the north-east border of Grand Teton National Park. It was around this time that Lisa started observing them from her plane. The following spring they became the Teton Pack when 24F gave birth to her first litter of five pups. Life was good: they had abundant food, healthy pups and were often seen together. However, no wolf has an easy life, and later that spring 133M got hit and killed by a truck on the highway, leaving 24F to fend for herself and her pups. Hunting elk for the whole family by herself was going to be a challenge.

In August she was found in a snare set to trap a grizzly bear, after raiding the grizzly's den in desperation for food. Miraculously she wasn't harmed by the snare, but the biologists who released her checked her condition and saw she wasn't doing well. The life of a single mother was taking its toll: her weight was just seventy pounds and her teeth were very damaged. It was decided by Grand Teton National Park that they would supplement the pack with food, so they were given roadkill of deer, bison, elk and moose to see them through to the fall when the elk hunt would leave her plenty of food to scavenge. The pups survived but, having got used to the food delivery service, 24F was seen feeding on a dead calf carcass that the rancher hadn't yet cleared, and no rancher wants a wolf around who has developed a taste for beef.

In 2000, with no mate, 24F didn't breed, but that September a new male, 137, joined their pack and became her alpha. Things were finally on the up again. Sadly, 24F and 137 never got the chance to breed. In November, Lisa was devastated to hear the mortality signal coming from 24F's collar – she was dead. They tried to get to her, but she was high in the hills above the elk refuge, and the snow was too deep and dangerous. Although everyone was keen to know how she had died, they had to leave her there. Eventually, when the snow melted and they could get to the body, they had it autopsied. No one expected the result that came back: she had been poisoned. They never found out who did it.

'She was a good wolf,' Lisa says, 'just minding her own business. By that, I mean that she never depredated cattle, just wildlife. It was just devastating, she was the first wolf here, and she was a great mum, a beautiful black wolf with golden eyes. I missed her, I'd see her at least once a week. I still get emotional about it. One of her daughters, 228F, became alpha and her blood is still around this valley, so it wasn't a wasted life.

'I understand that there have to be compromises, but we've all got to find those places where we can coexist. I can't believe we can't figure it out. We live in a modern world – everybody has to be willing to change a little, to take a step. Control measures should be used for wolves that are preying on livestock only as a last resort. Every non-lethal solution we can try we should try first. As far as trophy hunting goes? None of our wild animals should be hunted for trophies. I mean, people have to have that recreational opportunity, no matter what? That needs to change. I'm sorry, I just call it messy. All wildlife management is messy, but can we clean up the mess? I don't know.'

Just then we spot a grey wolf standing up in a group of willows near the new den. She is collared and luxuriates in stretching in the warmth of the low sun. We fall quiet now, just watching her through our scopes. After a few moments she ambles around, sniffing at bushes for a bit. She is in no hurry, just chilling. The burbling noise of cranes floats into the silent evening as they fly across the lake beneath us, golden in the light. The wolf begins to trot toward the new den.

'She could be the babysitter. I think she is young. Perhaps she can hear the pups in the den.'

She disappears. We wait, but there is no more sign of movement.

'So, what happened next?' I ask.

'Well, I carried on flying for a while but I retired when wolves were delisted. I just felt we were starting to go backwards. Then I found out about trapping and I started Wyoming Wolves Untrapped when suddenly wolves could be legally trapped again. I just thought it was the most depressing thing out there to bring back the wolves and then do that to them. That turned into Wyoming Untrapped. I had felt so close to this new family with all their troubles. I had watched them come back. 24F was the first wolf to give birth in our valley for over fifty years and this is what we do to them? 24F had a daughter who lived to eight and a half years old and had six litters. We got a mortality signal from her collar in 2008 but the genes of those black wolves live on. I guess many of the wolves here are descended from 24F.'

As the evening gets cooler around eight o'clock, another wolf gets up, but this time there is no ambling. This wolf is moving with purpose and wagging his tail. He is anticipating a greeting, perhaps. I keep the scope on him.

Sure enough, two other brown wolves step out of the willows he is heading towards. They are also wagging their tails. Then I realise that chasing behind the first wolf is another larger, darker one, and trailing behind that wolf are three pups. We are delighted to watch all the tail-wagging and greeting that ensues until they all end up crushing each other in a heap. Then, these descendants of the Teton Duo move off as a tribe, headed to the original den near a patch of trees up the gully. They travel so fast I can barely follow them with my scope.

We don't have much light left as the sky blushes behind the Tetons. The wolves play for a bit in and out of the gully before heading off up the hill. The last one I see is a small pup who nearly gets left behind while sniffing and digging at the bottom of a dead tree in the gully.

It is time to go, so I descend quietly. I have a head full of wolves.

CHAPTER TEN

A moment in time

'That den site you were at is still prime real estate for wolves. Now all the other packs compete for it. That is *the* spot for wolves.'

Steve Cain smiles under his bright white moustache. His deep voice is reassuring somehow, reminding me how comfortable he is with the knowledge he has gathered over twenty-five years of working for Grand Teton National Park's wildlife department.

We are sitting in one of my favourite places in Jackson, a sunny spot in the garden of Persephone Bakery, where they just happen to make the best coffee in town. The day is bright, the sun makes everyone smile, and the garden is busy full of locals as the tourist season hasn't really kicked in yet.

Steve was senior wildlife biologist for Grand Teton National Park when the den was first used by the Teton pack in 1998. He only retired recently, but he still remembers the first time wolves showed up in 1997.

'I was one of the first three people to see them back in '97,' he tells me as I get my own white moustache from the cappuccino froth. 'There was myself, Bob Schiller, my boss, and Mason Reid, another biologist. We got a call from someone to say that they thought they had seen wolves across the river from the main highway. We wanted to be sure it wasn't just coyotes so we dropped everything, packed our scopes and headed up to the Snake River overlook. It was fall – a beautiful day, I remember. We couldn't believe it. That moment, the picture I saw of the three of them through my scope is still etched on my mind so clearly even though it was a long time ago now. At that point none of them had collars and we had no frequencies anyway, so we didn't know who they were.'

'So it must have been the Soda Butte pack?'

'Almost certainly. We know that now.'

'As a wildlife ecologist, it was super-exciting to have them come down from Yellowstone so fast. In one way it was a surprise, but we know that wolves will travel hundreds and hundreds of miles, especially if they're searching for a mate, so we knew it was a possibility. It meant that we started developing the Grand Teton wolf programme right there and then. The following year they came back and settled and, as soon as we found out where they were denning, the park closed the area.

'Biologically, but also politically, the arrival of the wolves was a big deal. I think the park overreacted to a degree. There was a lot of law enforcement and an "incident management system" was set up – the superintendent wanted to prove that he had control. From that point of view, it was more about people than wolves, more about politics than biology. This was a super-sensitive area. Grand Teton didn't get the extra finances that Yellowstone did. We had to work collaboratively with Fish and Wildlife to ramp ourselves up with monitoring equipment – you know, collars and all that – so that we were in a position to monitor the recovery process and the population dynamics. That was important to us, that bought us a seat at the table for all the recovery discussions.'

'And the recovery obviously went well,' I say. 'Are all the wolf packs in the valley now descendants of that initial pack?'

'Well, that Teton Duo and the pack they started were the building blocks, but there were other immigrants too. By 2005 we were up to three packs, and then it really started to escalate. By 2007 we had seven packs, and by 2009 we were just shy of eighty wolves here. It isn't a slow process once they get settled – wolves can recover quite quickly.'

I wonder what those first years were like in that type of sensitive climate.

'At that time there was a cattle-grazing allotment near the den,' says Steve. 'That was the most difficult thing, and when 24F's mate got killed on the highway, we knew that those cattle might be at risk. It was a cow calf operation then, and that was one of the reasons the park made the really unusual decision to intervene. To do something artificial like feed the wolves with those roadkill carcasses was not something we would normally do in a National Park. It is really unusual, but it was mitigated somewhat by the fact that the wolves were on the endangered list and so we were in a good position to support them.

'Also, it is unusual to have a grazing allotment in a National Park – that was a historic thing. At first Grand Teton National Park just covered the area of the mountains, but in the 1950s it expanded, and at that point there were around twenty active cattle leases in the area. Suddenly Moose Head Ranch, for example, was part of the park but they still had rights to graze their cattle. Part of the expansion was making a deal with those ranchers that they could continue to graze their cattle and that their first heirs would also get those rights, but then it would end. So when the wolves came, there were three operators left who had rights to graze there. One of them was Cliff Hansen, who was also a governor and a senator. His grandson, Matt Hansen, ended up becoming governor of Wyoming, so now we are deep into the political arena.'

I think how it must have been pretty frustrating for those ranchers to suddenly have protected wolves on their doorstep, and so I ask Steve how it played out.

'The park put a lot of the onus to decrease conflicts on the rancher,' he replies. 'We were clear that state policy was that they would not remove a wolf in the park in response to livestock conflict but that the state reimbursement process applied. We advised ranchers, we gave them information about things like range riders, and to my knowledge, they did have a range rider for a while, but it soon stopped. I don't think any of those ranching families were happy.'

It's not hard to see their point of view.

I know the name Matt Hansen, actually Matt Hansen Mead. He is a Republican who worked as the US Attorney for Wyoming under George W. Bush. Less than a decade later he was campaigning to be governor for Wyoming, and during that campaign he was vocal about taking back control for Wyoming from the federal government. He was particularly active about the wolf issue. I suddenly realise it can be no coincidence that he holds resentment when he remembers his family were powerless to do anything about the new predators who were protected by the Endangered Species Act but were now living alongside their cattle. In 2003, five years after the wolves arrived, his family sold that ranch when his mother died in a horse-riding accident there.

While campaigning for governor, his press quotes were clear about his position: 'It's important for us to take control and be proactive to fend off federal regulation – wolves are symptomatic of this issue.'

'Sometimes,' he said, 'Wyoming is only viewed as a great big park with a great big fuel pump to be used by the rest of the country for the rest of the country.' And: 'Federal policymakers need to realise that we're not part of their park, they're part of our state.'

Mead was active in protecting grazing rights and supporting private property owners when they came up against those issues resulting from the Endangered Species Act. As governor, he would be staunch in his stand for the ranchers; his grandfather was nicknamed 'the cowboy governor', and it seemed that the same principles mattered to him. Mead described 'environmental extremists' as 'short-sighted', and said that if ranchers weren't able to use public grazing safely they would have to sell their land for development. He also campaigned for gun rights and anti-abortion laws and opposed gay marriage.

His campaign was successful and, in November of 2010, supported by Sarah Palin, he became the 32nd governor of Wyoming. He then began to talk about the success of the wolf reintroduction project and the increase in population size. After he had got that message across, Mead campaigned for Wyoming to be able to manage the wolf population independently because there was clearly no need for them to be on the endangered species list any more.

In late 2012 wolves in Wyoming were taken off the endangered species list. Much legal battle ensued, and they were put back on the list in 2014.

After years of wrangling in courts with pro-wolf conservation groups, on 25 April 2017, he was able to make the following announcement: 'I'm delighted that the Circuit Court recognised Wyoming's commitment to manage a recovered wolf population. Our wolf management plan is a result of years of hard work by people across Wyoming. We recognise the need to maintain a healthy wolf population… This is a good day for Wyoming.'

Was it a good day for Wyoming's wolves though? The argument was that the Wyoming wolf population was way over the target number established for taking wolves off the endangered species list that they'd had the honour of joining in 1974.

By October of 2017 Wyoming's wolf hunt season had begun.

Wyoming was divided into sections. No hunting would be permitted in either Yellowstone or Grand Teton National Park. Outside of these areas was a 'flexible management zone' called the 'trophy area', where only hunters with a licence could take a wolf and where there would be a yearly allowance for harvesting. Outside that zone, in the rest of Wyoming, wolves were classified as predators and could be shot on sight or trapped. This would mean nearly 90 percent of Wyoming wolves could be killed indiscriminately.

All this happened as Steve Cain was getting ready to retire.

'I think it's a shame,' says Steve, 'that Wyoming has chosen, for political reasons, to manage wolves at a minimum, when – as a wildlife ecologist – I can see that they contribute so much to the environment. I think they should be allowed to live at carrying capacity. They won't keep reproducing; every habitat has an upper limit, and there are only so many wolves that each place can ever support. But this is important – everything that lives in this landscape has evolved in the presence of wolves. We don't want to see those adaptive pressures change if we are to continue to enjoy this wilderness and all that it offers. And wolves help that. They keep down things like chronic wasting disease in the elk herd – which everyone is worrying about now – by taking out the weaker members of the herd. They are helping us to manage those types of things naturally.

'I think people really relate to wolves; they have a social structure, a pack structure. They sleep, move and feed as a group, and in that way, they are unique in North America among our large charismatic megafauna. But, although the political arena has changed since those early days, I'm not worried about them. Wolves will never go away now. I'm excited that they are part of the landscape. Over the last twenty or thirty years, I've seen the rewilding of Grand Teton National Park. When I first came here, there were barely any bald eagles, and they were still on the threatened species list, as were Peregrine falcons.'

I ask Steve what he feels about hunting and the anti-wolf brigade.

'Well, I'm a hunter, but I would never kill a large carnivore like a wolf. That is distasteful for me, but I have to accept that it is happening. Everyone has a different objection for a different reason, but it usually boils down to what it means for them and their family, their business. Hunters in Montana will tell you that elk numbers have gone down too far because there are too many wolves, and they will quote numbers and moan that they can't run their business. But I have a different theory about that,' he sighs.

I'm excited to hear this as I have always wondered how so few wolves could be responsible for taking the elk herd down so far. Steve is talking my language.

'Quite simply, I think that, although wolves were responsible for some of the reduction obviously, they weren't responsible for all of it. Hunters in Montana overdid it, especially the late-season female hunts, and that's why the elk numbers reduced so dramatically.'

'There seems to be such hatred for wolves,' I say. 'It seems to go beyond just being worried about the impact of a large carnivore in the neighbourhood.'

'There is a lot of blame, there is a lot of non-fact and a lot of misinformation. Before wolves came along, grizzly bears got the blame for everything. Then the wolves came back, and that blame shifted. I've seen it,' Steve laughs. 'I mean, your kid gets the measles? Blame the wolves. It was part of my job to speak on this stuff all the time. In recorded history, it's been shown that you are more likely to be attacked by a moose or a cougar than a wolf. There have been so few predations on humans in recent decades that they are noteworthy. One person in Alaska died, and three guys died after a wolf had bitten them, but they didn't die from the attack, they died from rabies.'

'What about this idea that wolves kill for sport and waste the meat?'

'Surplus killing, you mean? Yes, they will do that, but not for sport – the notion of them killing for sport is ridiculous – but because they are hunters that is how they survive, and in a natural situation they will come back to the carcass. There was a surplus kill in Bondurant south of here not so long ago – 2016, I think it was. The wolves killed nineteen elk in one night, which was rare, but what did they do? The people stacked up the carcasses by the roadside, so they never did get eaten. They wanted to make a point down there, to show all the people on the highway what the wolves had done. It would have been far better to leave them for the wolves to eat.'

'What about this idea that they reintroduced the wrong type of wolf, that this species is too big?'

'Again, that just doesn't make sense. We know that wolves are good dispersers. They will travel long distances to find mates, so there is "gene flow" all the way back and forth from the Rockies to Canada. If there had been an isolated population that had developed changes over a period of time and those wolves were selected from that, then it would be a fair criticism. But these wolves were selected from the general population.'

Finally, I ask Steve about the future, about what he thinks will happen next.

'I'm really looking forward to the day we analyse the genes,' he says. 'I think we will find there are complex relational factors governing behaviours that we don't even know about yet. For example, there are two cases of grizzly bear adoption that I know of during my time at the park – one with 399 – and in both cases, the daughter took over the care of her mother's infant. We know those families and those individual bears. We have never seen another case of adoption like that, and I'm guessing that non-related individuals might not do that sort of thing. So there is probably a lot of behaviour in wolves that, when we know more about how they are related, might be set in a different context. That is exciting for me as we come into this new era of science and research.'

CHAPTER ELEVEN

I'm just another groupie

As I drive toward the austere Roosevelt Arch guarding the north entrance to Yellowstone National Park, the light is already fading. I find the fifty-foot arch of dark columnar basalt foreboding rather than impressive. Something about its complete incongruity with the landscape gives me the shivers and says nothing of the delights of Yellowstone National Park to come.

The US army, drafted in to create the roads and buildings of the park, built the arch in 1903. Roosevelt himself attended a grand commemoration ceremony to lay the cornerstone. He said, 'Nowhere else in any civilised country is there to be found such a tract of veritable wonderland made accessible to all visitors, where at the same time not only the scenery of the wilderness but the wild creatures of the Park are scrupulously preserved, as they were, the only change being that these same wild creatures have been so carefully protected as to show a literally astonishing tameness. The creation and preservation of such a great national playground in the interests of our people as a whole is a credit to the nation; but above all a credit to Montana, Wyoming and Idaho.'

The evening light is still bright enough for me to read the inscription: 'For the benefit and enjoyment of the people.' Driving underneath the arch, I am made to understand that the creation of the park was for the people, and the 'protection of the animals' was as part of a playground for the people, not for the animals themselves.

Things have changed since then. In 1995, wild wolves were reintroduced to Yellowstone, in one of the most controversial, famous and successful conservation projects in the world. This is why I'm here — simply to be what I have always been: a wolf groupie in the best place in the world for it.

The lady at the modern north gate's entrance hut slides open her window and is far more welcoming than the Roosevelt Arch. She smiles under her iconic National Park uniform brimmed hat, comments on how cute my dog is sitting in the passenger seat beside me and gives him a treat — a habit at these entrance huts. My Goldendoodle — predictably named Doodle — is very welcome in the park, but he won't be getting out of the car tonight as he probably wouldn't last long with the pack of wolves I am hoping to see.

Snow is on the road already, and there is more forecast, so I'm keen to get as much of the drive done in daylight as possible. I am headed to see my friend Karen in Silver Gate, just beyond the north-east edge of the park. It is as close to the famous Lamar Valley, home of wolf introduction, as it is possible to get without staying in the park.

Silver Gate is a tiny place, just a collection of cabins really, plus one very large building called the Range Rider. This barn-like building has always fascinated me, and not just because of its name. It is a vast, double-storey log cabin, entirely out of scale with the rest of Silver Gate. When I asked, I learned that it was a dance hall. I once got the chance to sneak inside and take a walk around its dusty, languid interior with Karen and our kids. Lining the upstairs balconies are small bedrooms, each inscribed with the name of a girl: Lilly, Tilly, Milly — you get the idea.

'Dance hall, my arse,' I found myself calling through the echoing space to Karen. 'I reckon the Range Rider might have been a whorehouse.'

While the kids ran around downstairs, I couldn't help opening a door or two and peering inside, all the while fighting a suspicion that I could be disturbing an intimate moment. Each small bedroom had the exact same furnishings. An old-fashioned, single, wooden bed and a dresser with a plain white pitcher and bowl on it – a scene right out of a western movie. Downstairs I even find swinging saloon doors, which I have to walk through a couple of times to get the right dramatic effect.

Exploring one of the side rooms, I discovered that apparently, the great Ernest Hemingway was a 'regular' at the Range Rider in the 1930s. Along with a fishing rod, there was a collection of other items he may have left behind. Some of his work is proudly exhibited, set in a typewriter as though he had just taken a break from writing to pop upstairs 'for a bit'. Although apparently, he did stay here with his wife.

I couldn't help but hear the noises that would fill this vast building – the parties and people that it must have seen. So far into the wilderness, removed from civilisation, it is easy to imagine that the illegal activities that must have gone on here didn't really come with much consequence.

Sitting primly next to the Range Rider and beneath the curves of Amphitheater Mountain are the pretty little cabins I am headed for today in Silver Gate – a name that always conjures a magical picture of stars with a full moon bouncing off the crags of the mountains, silhouetting the conifers and making the snow glisten. I am excited to get there.

My way takes me through Mammoth, home of the National Park offices and some employees. It is an odd collection of army-built, utilitarian park architecture that bustles every summer with tourists checking in for lunch after working up a hunger exclaiming over the beauty of the Mammoth Hot Springs terraces just up the road. Mammoth is an odd place, retaining the sense of being the last stop before the wilderness. It even has its own small jail should anyone be caught misbehaving. Here, the lawns are kept to a neat length by wild elk. My favourite part about it, though, is that this is where the park scientists work. The first time I came here to meet them, I felt a bit like I first felt when entering BBC Television Centre – a little giddy. Ecologists, biologists, the wolf conservation team… here is where the magic happens.

This evening, devoid of tourists, Mammoth seems deserted. I turn left and out into the park, heading over the bridge that spans a huge canyon, and now I really feel as though I am in Yellowstone. Only one more left turn at Tower Junction and I am nearly there – well, around an hour away, but everything is bigger in America. At this time of year I can't get lost, even in the dark; because of the deep winter snow, these are the only two roads open.

The night is black as I finally enter the temple of the wolves – Lamar Valley – the place of their reintroduction, the setting for the soap opera of their lives. I'm listening to the audiobook *Wolf Nation* by Brenda Peterson. It is an intricate blend of science and storytelling and, having driven for over six hours to get here, I am on chapter seven. It doesn't feel like any coincidence that I am listening to this chapter here, now. I may not be able to see anything at all but the narrow, grey road ahead and the swirling snow in my headlights, but I know that all around me is vast wilderness and somewhere in that wilderness are wolves.

Driving through the darkness, I learn the story of a wolf called 06 – a rock star here in the Lamar Valley. Her real name was 832F, but her nickname came from the year she was born, 2006. As I listen, I picture her in my mind's eye. A large, dark wolf born to the Agate family, 06 became famous as the charismatic and strong matriarch of the Lamar Canyon pack. Known as the 'legend of Lamar Valley', this strong female was a survivor. After her father, the alpha male of the Agate pack, died, she left. As a single, female wolf, she fed herself alone through a whole Yellowstone winter, against the odds and despite concerns from the wolf watchers that she would never make it.

Wolf Nation describes tales of her prowess in the hunt, even though she was mother to many pups, and of her bravery, especially in territorial combats with the famous and intimidating Mollies pack. In one story she survived an attack alone against sixteen of their wolves simply because she was clever: she foiled them by leading the wolves to a cliff edge and then disappearing down a hidden path they didn't know about, making her way safely back to her den. At the peak of her reign, she had nurtured and fought for the Lamar Canyon pack until it was thriving and a favourite of the many wolf watchers who came from all over the world.

At some point soon I will cross the border into Montana, but in the darkness it is impossible to know precisely where. For now, it is enough to know that I am in Yellowstone – where any wolf is protected. As I leave Yellowstone, that will change. In 2011 wolves were taken off the endangered species list in Montana and legal wolf hunts were sanctioned.

Finally, the shadows of buildings come into sight, and I breathe a sigh of relief. I'm at the north-eastern National Park gateway. No one is operating it at this time of year, so I can just drive on through. My shoulders ache, Doodle is bored, and the snow is really driving down now, blurring my vision. The windscreen wipers can only just about keep up with sweeping it away, and my max speed is just 20mph. I calculate that I am certainly in Montana now but, more to the point, I am almost in Silver Gate. Driving out of the park, even in the snow and darkness, this border between protection and danger is clear to me but remains invisible to any wolf.

In late 2012, as the fall elk were getting thin on the ground, the Lamar Canyon pack needed to roam further afield to hunt. Unwittingly leaving the protection of the park, they crossed the invisible border into Montana. The hunters were ready. The first to be shot was 754 – brother to 06's alpha mate, uncle to countless pups, partner in so many hunts. His absence was a huge loss to the pack but not significant in terms of its structure.

That December, a hunter – who, knowing the consequences, chose to maintain his anonymity – shot 06. After years in the making, the legend of the Lamar Valley was lost in an instant. The expression of grief was worldwide, with her death even making headlines in the *New York Times*.

The snow-laden trees around me disappear as I arrive in Silver Gate. The clouds are clearing, and the almost full moon shines on the fresh snow. With a sigh of relief, I crunch to a halt in front of the brightly lit cabin. My friend Karen, having seen the headlights, is already opening the front door, spilling warm yellow light into the night, but tears shine on my cheeks. How losing such a strong leader would affect the Lamar Canyon Pack no one could predict.

That night we hear 06's ancestors, their howls echoing around the amphitheatre. The Lamar Canyon pack – now just three adults and four pups – are once again out of the park, over the invisible border between protection and persecution. The next day we might get more of an idea why.

CHAPTER TWELVE

The soap opera

I wake to an early alarm. This doesn't usually inspire me, but today I am immediately excited. This morning has a feel of Christmas Day about it – you never really know what you will get. We will be wolf watching in Lamar Valley, getting to know the soap opera that is wolf life here. Like any soap opera, this one is dramatic – its plotline features life, death, elk kills, etc. – but, unlike any soap opera I have ever watched, there is no lounging on the sofa with a glass of wine; to view today's episode means getting up well before dawn.

I hear Karen's alarm in the next room and a subsequent groan. Outside it is still dark as I fumble around in my bag, finding layer after thermal layer, sock after thermal sock. A few minutes later, wrapped in thick boots, scarves and icy darkness, we slip and slide down the wooden steps from the cabin. The previous night's drive has left ice formations on my car that I can only compare to lichen. The overnight temperature has frozen the doors shut but, after a brief wrestling match, we yank them open and clamber in, Michelin Man-style, our plumes of breath steaming up the windows. We clasp thermal mugs of hot coffee as if our lives depend on them.

The temperature reads -25°C. It is all rather grim but for the tingle of excitement we both feel at the thought of seeing wild wolves. I turn the car towards the Lamar Valley.

Light creeps into the sky and our world becomes faintly visible. We pass the dark shape of a bison and then stop to watch two moose picking their way over the snow near the Soda Butte Creek, which runs parallel to the road. The creek itself is hard to make out as the ribbon of grey water slips underneath the puffy white snow overflowing from the banks like shaving cream.

The moose speed up a little when they see our car slow down, but they soon realise we aren't coming closer and they calmly step off into the trees. Like us, they are headed for the park, a place where wildlife is protected, where moose can be moose, and where wolves can be wolves and eat them.

The sun slowly rises, revealing the Lamar Valley – it is mystical, breathtaking. Steam lifts from streams banked in blankets of pure white, cottonwoods are coated in thick hoar frost – they seem to shiver with relief in the sun's low warmth. The smaller willows below them in the shade seem to find it harder to bear the weight of the ice. That may well be where the moose are heading to feed; they like willows.

Another bison, its dark fur ice-encrusted and its tiny black eyes closed, lies hunched in the snow horizontal to the rising sun, waiting to absorb every last ray with his thick dark coat. We stop to photograph him. The thick frost hasn't discriminated between him and his surroundings; it is as though he has a halo of white between his two small horns. That same frost covers every leaf of the sagebrush around him. He would only have had the protection of this coat through the night. It is hard to see how any creature could survive out here.

Everything shimmers. It is so cold that ice crystals float in the air, and each rising grass head is a flower of sparkles. We are in another world.

We drive slowly so that I can look for otters in the warm waters of the Lamar River. I see plenty of slides where they have slipped down the banks on their bellies but no actual otters. The banks are covered with trails carved from different feet. We spot elk tracks and possibly snowshoe hare.

The valley broadens, almost as though it is a large stage with the road at one side, perfect for an audience. We drive past the confluence of Soda Butte Creek and the Lamar River. The rivers combine in a rush and the Lamar takes the water from here, busier, more musical. Over time, its meandering has carved the broad plain to our left, and the far rising slopes are peppered with groups of conifers and large flat rocks. Behind them rear the mountains. To our right, the slope is steep, cliffy and more wooded.

We round a corner and immediately spot a lay-by – or turn-out, as it is known here – full of people: wolf watchers. My immediate reaction is to be intimidated; these people know what they are doing, they are dedicated and knowledgeable. I wonder if they will resent us amateurs for turning up, taking valuable scope space and asking ridiculous questions. At least ten people wrapped in multiple layers are bent over scopes on long-legged tripods, each aimed in precisely the same direction across the valley. Here are our wolf watchers and they sure are watching something. I realise I'm just as fascinated by getting to know the wolf watchers as I am excited about seeing the wolves.

At the far end of the turn-out, we discover just one, very narrow, car-parking space remains. Immediately a guy, so cocooned in layers that only his sparkling blue eyes are visible, waves us in. We are already welcome.

I open the back of the car and bring out the scope. Kit has always been my husband's territory, but now this is my adventure, my kit. In my thick gloves I fumble with the tripod but the lovely Chris, who guided us into the parking space, leaps to my assistance: 'Let me set it up for you. I know where they are. It will be a lot quicker.'

I am grateful, and while he does that I search around for my notebook and trusty fountain pen so that I can have them handy in my pocket. Chris explains that he is a guide working out of Livingston and his is the minibus we are parked next to. He introduces us to his Japanese guests, who are taking turns to get out of the minibus, have a look

through the scope for a few seconds, smile kindly, nod
politely and get back in the van.

I notice one of them who must be struggling. This
guest clearly didn't get the memo about the temperature-
appropriate clothing, and he is just wearing a rather fetching
hoodie, fashion-forward jeans and immaculate trainers. His
smile is rather forced – well, it's hard to smile when your
teeth are chattering. Chris, looking up from my now
perfectly focused scope, catches my eye.

'Nothing I can do, I have spare coats in the van, but he
won't wear anything else.'

We shake our heads in unison. Busloads of tourists
from Asia have become more and more familiar both in
Yellowstone and Grand Teton Parks in recent years. One
beautiful summer day, I remember sitting at Jackson Dam,
overlooking the lake and the Teton mountains. It is a
beautiful spot, and the children were contentedly skimming
stones. I was glad to have all the time in the world to drink
in the view and breathe the pure air, and I was filled with a
huge sense of gratitude to have the privilege of becoming
more and more familiar with this special place. I counted
four busloads of tourists arrive, take their photos and
depart in just the time I sat there. They barely had time to
taste the air. Now it appears that even the rare sight of
wolves isn't enough to make them linger.

'Take a look,' Chris says, indicating the scope.

I can't help but gasp at the scene that comes into focus as
I place my eye to it. After many years of working in wildlife
film-making, I understand that animals rarely – if ever –
turn up on cue, but we have lucked out on our first morning.

On the far butte, in front of a group of conifers and one
dead tree, the land levels. I count seven wolves feeding on
something bulky lying in the snow, surrounded by ravens
and magpies all eager to sneak a share. It is impossible to
even guess what they are eating. One black wolf, having

filled his belly, wanders slowly away for a lie-down, clearly feeling a bit like my husband after a big Sunday lunch.

'They're on a bison carcass right now,' Chris says.

I pull away from the scope to let Karen have a look. Now I know where they are, I can just about pick the wolves out with my naked eye. They are as tiny as ants. Time to begin writing notes.

'Oh, they are fantastic!' Karen breathes, her words disappearing into a fog of warm air.

'Did they kill it?' I ask, feeling like a journalist as I whip out my fountain pen.

'We don't know. They were on it when we arrived before dawn. The biologists down the other end of the turn-out will have a better idea.'

Chris tells us this is the Junction Butte pack as I peer into the scope again and say, 'I can see seven.' He tells me to move my scope to the right. 'Oh! I see another two!' Two black wolves are playing a small distance from the carcass.

'There are actually eleven,' Chris continues. 'The alpha male and female have already eaten and are taking a nap. And there is another wolf curled up in the snow just above where those two are playing.'

'Oh, I see him.' I pull away from the scope to write notes and learn my first important lesson when writing about wolves: a fountain pen will freeze in -25°C, rendering it utterly useless. I find it rather embarrassing as a writer to have to ask to borrow a pen. Still, nothing is too much trouble around here, and Chris whips one out of the minibus in no time.

There is nothing more fun than watching wolves play, even when it is so cold. It doesn't make any difference to your numbing feet, but it does warm the cockles of your heart.

One by one, the younger wolves leave the carcass. They are this year's pups. You can't really tell this from their size because they are pretty much fully grown, but you can certainly tell from their behaviour. While the older wolves want nothing more than a digestion break, the youngsters – fuelled by their bison feast – are full of energy and so tease each other into a game. This is the moment the birds have been waiting for: they pile onto the carcass, pulling and cawing, but the wolves don't really care any more. They have found a tall, flat rock and launch into a game of King of the Castle on it. Up then down and up again – there doesn't seem to be much sense in it, it's just a way of expending energy. Then they wrestle, leaping with long straight legs into a tail-wagging downward dog, posturing over and again before launching into full attack. They circle and chase each other, pushing one then the other down into the snow, jaws closing gently around throats and tails. I find it impossible to look away, I don't want to miss a second. Two of the adult wolves, torn between sleeping off their meal and playing, finally admit defeat and – bums in the air, tails wagging – they join in the game, all coming together until five of them are in a heap.

One of the black pups revisits the carcass. Irritated by the corvids that have covered it, the pup rushes at it, chasing them off. He is obviously keen to conserve some of the meat for snacks later. After charging the birds he enthusiastically rushes back to rejoin the game.

Behind me, I hear Chris announce to the full minibus: 'Well, if no one wants to watch the pups play, then we can head out for breakfast. Don't feel that you have to stay; it's not an endurance test.' I look up to see his passengers nod enthusiastically at the thought of breakfast.

When they have gone, we move our scope to the other end of the turn-out and introduce ourselves to the biologists. One is doing push-ups to get the blood flowing again before he peers back through the scope. Claire, a student with the wolf project in Yellowstone, turns from her scope, smiling. While I set mine up next to hers and peer into it, she chats to me.

'So,' she explains, 'each wolf has a number. Right now we are looking at 907 – the grey one with the collar. She is the alpha female. 969 is that grey subordinate female, and the other is 907.'

I can just make out the thick black collar on the alpha female. The wolf project collars wolves each year and the ideal situation is for them to have a collared wolf in each pack so that they can keep track of their movements, but that doesn't always happen.

'Junction Butte pack currently consists of eight adults and three pups,' Claire continues, 'so the two grey ones you can see are both adult females. I think the alpha male is asleep back there in the trees. This is a pretty old pack. I think it was established in 2012?' She looks to the two men who nod their approval. 'But they are doing really well at the moment, they are strong. They have a den, actually, in Slough Creek,' she says, pronouncing it Slew – an odd Americanism. 'It's quite a few miles east of here, but they came up to Lamar Valley in the night, and this is where we found them this morning. This is often where we would expect to find the Lamar Canyon pack, especially since there is meat on the ground, but the Junction Butte pack are stronger than them right now. The Canyon pack have five pups and fewer adults – only three – so they won't risk a fight with these guys. They would rather move on than confront them. The pups are pretty much full grown but they haven't had much life experience yet, and they wouldn't be much use in a fight. Most people don't realise that pups have to learn to fight. Fighting isn't something that is purely instinctive. It will come – territorial disputes are a normal part of wolf life – but for now the Lamar Canyon pack have gone further east than normal, into Silver Gate out of the park.'

'We're staying there. We heard them last night,' I mention.

969 wanders away from the group to curl up nose to tail in the snow and sleep. I can barely talk because my lips are so numb with cold, and we all have drips on the end of our noses. I don't mind, though. I really feel that qualifies me to join the hardcore wolf watchers.

'How can they possibly go to sleep out there?' I ask.

'They seem to be able to sleep in any conditions,' Clare laughs. 'Seriously, we can be here in below freezing temperatures with the wind blowing so hard we can barely keep our scopes standing, and they are sleeping as if it was a warm summer day. They are built for it. It is amazing how tough they are.'

'We're not sure if they actually killed the bison or not,' mumbles the young guy next to her from underneath his balaclava. 'It doesn't seem to be that fresh. If it were, there would be bloodstains all over the snow, and if you look at it closely, you can see that it is pretty hard to distinguish what it is. There isn't really much left of it.'

We hear the drone of a small plane, and a few minutes later a small, silver aircraft comes into sight, flying low up the valley. The biologist's radios crackle and I hear a familiar voice: Kira Cassidy, head of the wolf research team. I will be with her tomorrow when she supervises each of the groups, but today she is coping with the cold in the cockpit of a Cessna. At this time of the year, the biologists on the ground and in the air are here because they are working together, taking a census and doing some collaring. Most days Kira or even Doug Smith (leader of the 1995 reintroduction project, and something of a legend in the wolf world) will be up in the air, taking masses of photos that they will use to make a count of the wolves and observing the movements of each pack.

The plane passes low over the pack a few times, as I wonder how much different the view of them is from the air. The wolves don't even look up. They are used to the planes and know that they aren't a threat. Any time they are darted and collared, it is done from a helicopter, so Claire tells us that they are much more wary of those.

From the radio, Kira's voice comments that they have everything they need and that they got some great photos. Then she is off into the deep blue sky and over the butte to find the next pack.

On the ground, one by one, the wolves give up their game and settle down for a nap. If I were them, I would be curling up together for warmth, but they choose to sleep separately, dotted across the slope.

Show over. And yet the biologists keep watching.

'Now is the time when another predator might make an appearance at the carcass.'

'Haven't all the grizzlies gone to bed?' I ask, slightly nervous that it's an amateur's question.

'Well, the sows with cubs will be in bed for the winter by now,' mumbles balaclava man, 'but just yesterday we watched a big male wander all down the valley, so we are curious to see if he is still around and if he might make an attempt on this carcass. He could well be waiting in the trees back there.'

'What will the wolves do now?' I ask.

'They'll probably sleep for most of the day.'

Karen and I look at each other, and I know we are both thinking of hot coffee and breakfast.

'So what will you guys do?'

'We have to be here. We have to monitor every move, who does what when. We don't get to leave when they sleep, we have to just wait and count the hours.'

I really feel for them. I am not so dedicated. The cold is becoming painful now. My fingers are in agony, and I wonder if I will ever be able to read the notes I'm making. Every exposed piece of skin hurts, and I can no longer feel my feet even though I have thermal socks and shearling-lined snow boots on. I can see that Karen feels the same way. While we are reluctant to leave, the lure of warm coffee and something to eat is becoming too much to resist.

'I think they are still serving coffee and breakfast at the hotel in Cooke City,' she says quietly.

I have a whole week of wolf watching ahead of me.

As we prepare to leave, another regular shows up. The biologists exchange warm greetings with her. In this game, there is little distinction between the dedicated wolf watchers and the biologists, as all information about the wolves – every small observation – is valuable in building a picture of their lives and interactions. Fresh from the heat of her car, she introduces herself with a warm smile as Melba Coleman.

She has been with the Lamar Canyon pack over in Silver Gate several times in the last week and shows us pictures on her phone of 926, also known as Spitfire. I immediately recognise that number – this is the daughter of the legendary 06 that I heard about as I drove through the park last night. The photos are taken at really close range and show a black wolf with grey on her chest and muzzle. She is on the road, walking through Silver Gate with her pack.

Melba tells us that 926 led the pack for seven or eight years until last year, when her own daughter – around three years old and known as Little T – displaced her as alpha. Little T's alpha male is Small Dot, named for the small white dot on his chest. He is unusual in that he was born outside the park to the Beartooth pack coming to Yellowstone in 2016. Leaving the natal pack to try and join another one is always a high risk for a wolf, but it is crucial behaviour because it prevents inbreeding. For many wolves it is a move that costs them their life – it is hard for a wolf to survive alone, and when he does come across another pack, they may well kill him. But for Small Dot it worked out just great: not only did he join the Lamar Canyon pack but he became alpha male.

Seeing how interested we are, Melba starts looking for some footage on her phone.

'You'll love this,' she says. 'Listen to her howl, she actually has a wolf whistle.'

We watch the wolf howl and, sure enough, just before the low howl begins we clearly hear a whistle.

'She does it every time,' Melba says, laughing at our reaction. 'Wanna hear it again?'

We do – it is hilarious: a little whistle before every howl.

'We think it might be something to do with the configuration of her teeth, but we don't really know. Anyway, don't tell anyone. We want to keep it as quiet as possible because it makes her so distinguishable and then she's an easy target for hunters.'

Like many of the wolf watchers, Melba is an authority on the Yellowstone wolves, and she is involved with an online group that maps the complicated genealogy and family trees of each individual and each pack. It is an impressive project. I am only just beginning to get a handle of the complex web of relationships that make up a wolf pack and the family lineages that now play out life after life on Yellowstone's famous stage.

We won't say a word we promise, and I fully intended not to include it in this book.

A short while later we drive through Silver Gate and almost immediately we are into Cooke City, which is as far from being an actual city as it is possible to get. To be honest, you would struggle to describe it as a town; it's more like a very, very small village – no, make that a hamlet. It's basically a collection of wooden buildings clustered together along the main street with a couple of petrol stations plonked in the middle. Today it is bleak and the streets are white, the few shops and cafes that serve the tourists on the doorstep to Yellowstone in the summer are now closed, including a wooden shack displaying a sign that randomly offers fresh Asian food – not something I would have expected to find in the wilds of Montana, but I wouldn't have been able to resist had it been open. Today, the only sign of activity is the distant drone of the snowmobiles, a sport for which this place is famous in the winter. Snowmobiles for rent line the forecourt of the Exxon petrol station, and we pull in and park next door.

I have my doubts that Soda Butte Lodge is actually serving breakfast. A massive fire burns in the foyer, but the place seems deserted. Within a minute, though, a slight Indian man appears from what looks like the dining room. He is very happy to see us.

'Might you be serving breakfast? Or even just a coffee?'

'Oh! Yes,' he beams. 'Come to the dining room, we can certainly do that.'

He ushers us to a table by the window and pours hot, bitter coffee into brown mugs as he explains that he is the new owner. It is clear that we are the only customers, and I suspect that we might be the most exciting thing that has happened in Soda Butte Lodge in a couple of weeks at this time of year. While we wait for breakfast, he insists on giving us a quick tour of his new pride and joy, especially the bar with its curved windows looking out on the creek. He is thrilled at our compliments, and as we head back to the dining room, even the cleaners come to take a look at us – real customers.

Breakfast is served, and with it, the new chef, Keith, comes to join the chat. As soon as we explain what we have been up to, he rummages in his deep apron pocket, then whips out an iPhone and begins flicking through his photos and we see – guess who? Spitfire, 926. The shots are close – very close, she fills the frame – yet Keith is no wildlife expert or photographer. There was no telephoto lens involved, Keith used nothing but his iPhone.

'I've never seen anything like it,' he says, beaming. 'My first week here and I am this close to a wolf. Nothing like that has ever happened to me before.'

We try to be polite and hold off eating while he describes it, but we are starving, and melted cheese is oozing out of our omelette. Keith is so excited about his new-found love of wolves that he clearly isn't going to stop raving about his experience, so we delicately dig in and nod and smile with our mouths full. Soon there is no pretence at small mouthfuls and, while we are busy scoffing the whole plate of omelette and hash browns, he shows us a mini movie of Spitfire walking right by his car. Delight is written all over his face; he obviously still hasn't quite got over the experience and probably never will.

'I gather they are still in Silver Gate,' I say. 'The Junction Butte pack seems to have been pushing them east.'

'Oh, they're around all right,' says Keith with a big grin. 'Isn't it amazing? Wolves on our doorstep. No wonder people want to come and stay here. It may be a lot of snowmobilers here at the moment, but there's a heck of a lot of people come to see those wolves too.'

The new owner nods happily. 'We will rely on them,' he says.

When we get back to the cabin, we are met with two sets of pleading eyes: Doodle and his best mate, Karen's tiny rescue terrier, Halo. So, despite an increasing post-brunch need to just loll about and doze in front of the fire, we put our layers back on, giving in to our dogs' desperation for a walk.

Chatting about the morning adventure, we stroll between snow-laden conifers and cabins along a back road of Silver Gate. It is so little used that it wasn't ploughed even after all the snow last night and our path is still fresh and white.

Even though he is hanging out with Halo, I notice that Doodle is distracted. He keeps stopping to stare into the dark woods and sniff the air. We know the wolves are here somewhere and I can't help wondering what he is so fascinated by.

Karen reads my mind: 'Do you think they could be holed up in there for the day?'

'I guess. They could be anywhere.'

And then I see the tracks.

'Look!'

There is no doubt they are wolf tracks; they are larger than any dog's. Both Karen and I have seen wolf tracks before and we are thrilled to see them now. They wander up the road in the same direction we are headed. The fact that they are in fresh snow means they were probably made in the early hours of the morning. We are literally walking in the footsteps of the famous 926, Little T, Small Dot, and the pups that make up the rest of their diminished but hopeful pack. It feels great. I put Doodle on his lead, just to be safe, so he is less than happy about the situation, but I remain deeply grateful to be so close to wild wolves.

That night, after a couple of glasses of wine, Karen and I sit outside on the porch and howl into the night. Our voices echo around the high rocks of the amphitheatre and, holding our breath, we wait… but the air is still, there is no answering call. The wolves are either laughing at our pathetic attempt at their majestic howl, or they have moved on. Perhaps, over the coming week, I might find out.

CHAPTER THIRTEEN

Elk curry and wolf talk

The photos that Kira Cassidy shot from the small, silver aeroplane are nothing short of spectacular. Despite many years of wolf watching and being the leader of the project, she is as excited as I am as we sit at her kitchen table and watch the pictures load onto her computer.

I have moved from Silver Gate to stay with Kira in her cabin in Gardiner, and we are surrounded by lolling dogs and wolf paraphernalia. Huge books about natural history and science overflow from the shelves and sit in piles on the floor. The small kitchen has everything we need to make food, although nothing matches. An oversized, old wood-burner pumps out heat, so the living room is warm and comfortable after a long day, while in the cold mudroom a half-butchered, road-killed deer has not been forgotten by the dogs. It strikes me that this is far from the environment of the domestic goddess that I have been aiming for throughout my married life with children. This is the environment of a woman following her passion – a wolf goddess.

Kira is small and slight yet utterly determined. Her passion for wolves brought her to Yellowstone, and now she is a research associate with the Yellowstone wolf project. She is especially interested in wolf pack behaviour, sociality and territoriality.

We each have a steaming bowl of elk curry, yet all we're really interested in are the stories coming from her images. Yesterday didn't seem to bring much action on the ground for the Junction Butte pack – they had left the bison carcass and moved on, so there was nothing to see from the road but plenty to see from the air. And man, did they cover some ground. The aerial shots show them running together, tails raised. Kira explains that this body language shows they aren't being aggressive, although they are deliberately covering the ground right into the Crevice group's territory and are possibly on their scent.

'You can see how confident they are. They feel like a strong pack at the moment.'

I thought back to yesterday's comments that the Junction Butte pack were strong enough to send the Lamar Canyon pack east out of the Lamar Valley and out of the park without even having to fight them. They are clearly on the rise.

More of Kira's images, also taken from the plane, show the wolves fording a rocky part of the Hellroaring Creek deep in a canyon – something no one could have seen from the ground. The series of shots reveal the adults going before the far more cautious pups, turning back to encourage them on. Finally, just before Kira left them, they reached the frozen pond on the Crevice pack's territory, where they ran around excitedly, sniffing and marking.

Together, next morning, Kira and I drive through another blue-sky, hoar-frost morning into Yellowstone Park. Kira's job today – rather than keeping track of wolves – is to keep track of the teams of biologists who are following them.

'Little by little,' she tells me, 'at this time of year, between the flights and the teams of biologists we build a picture of wolf behaviour, of pack interactions, of where they go and perhaps why. Every single observation is useful. Over many years of this kind of work, we've gotten an understanding of the complex relationship dynamics between the groups.'

We dip into the early-morning briefing session in the wolf project office at Mammoth. I could have lingered there all day, but there is no time and we have to hit the road.

As she drives, Kira continuously checks her radio. First, we hear from the researchers on the trail of the Eight Mile pack subgroup: 'We don't have any visuals, but we do have a clear radio collar signal from…' and then it stops making sense, as if they are speaking in a foreign language.

A certain look settles on Kira's face for a second – a mix of serious and concerned. Then she turns to explain.

'The wolves are headed up on North Butte. Depending which direction they go, they could cross the boundary and leave the safety of the park. Hunters are waiting for that. Often our radio comms are being scanned by hunters, so we have to be careful what we say and talk in code.'

We are silent, watching the day's early sunshine flicker between the trees as we drive. The radio comes to life again, and this time it is project leader, Doug Smith, who is in the plane today.

'The 962 subgroup – four black wolves and one grey – are up Everest where no one else can spot them. The wolves are on the move so the ground team might as well take a break – they won't see them for at least a few hours.'

He also says that one group of wolves are 'for sure' out of the park and that it might be worth checking for them in Gardiner.

'This is a bad situation for the wolves,' says Kira after Doug has signed off. 'The minute they are out of the park their lives are at risk.'

We pull into Hellroaring Overlook, a turn-out where our first group of warmly wrapped biologists are waiting. They have been here since sun up. They flap their arms and stamp their feet in the continual effort to keep warm that I am still getting used to. The icy temperatures are a major downside of the job but don't seem to sap any one's enthusiasm, and they are all smiling.

The deep Yellowstone River valley stretches out in front of us, drinking in the sunshine. Conifers clad the slopes below. On the other side rise mountains and hills, part-covered in patchy snow, and between them and us, Hellroaring Creek joins the Yellowstone River.

This team are dedicated to the Crevice pack, and they start filling Kira in on yesterday's observations. It transpires that yesterday the Crevice pack were actually playing on the frozen pond in their own territory not twenty minutes before the Junction Butte pack arrived on it. Slowly I am beginning to see that every observation really does count towards telling the story of exactly what is going on for Yellowstone's wolves, the power struggles, the dangers.

'Ah! That explains why Junction Butte were so excited when they got there,' says Kira, sharing her observations from the air. 'They must have smelled the Crevice pack and knew how fresh the scents were.'

Of course, working in a place like this provides the chance to see lots of wildlife, and it isn't just the wolf stories that are interesting. The team are excited to tell us about some other action that played out through the course of yesterday afternoon. From their position, they were able to watch a large bull elk, alone in the valley below, face terrible odds. First a mountain lion stalked him and then tried to attack him, which is pretty rare for anyone to see, even in Yellowstone, but the elk heard him and swung around, lowering his head and brandishing his antlers. Quite quickly, having lost the element of surprise and realising this would not be an easy meal but possibly a dangerous one, the cougar gave up.

But the excitement didn't end there. Through the rest of the afternoon, the team were impressed as the bull elk stood his ground against not one, or even two, but three grizzly bear attacks. The biologists were amazed that not only had the elk survived four attacks in a single day but that he still didn't have a graze on him by the time they left at nightfall.

One by one, while we talk, we notice a sweet female elk attempting to tiptoe past us unseen, slowly making her way around the edge of the turn-out. Every time we look directly at her, she stops moving, her big brown eyes watching us, uncertain – should she be afraid of us? So we give her the blessing of ignoring her and, in the manner of a little old lady in a rush to get the bus, she crosses the road, clambers up the bank and then happily disappears into the forest.

We learn that there are apparently seven wolves missing from the Eight Mile pack so that pack is currently split into two, but no one really knows why or where they are. Then we hear wolf howls echo across the valley. Might that be one of the groups trying to communicate with the other? All eyes are immediately down to the scopes, but there is nothing to see.

Reluctantly Kira and I get back into the car, although we are glad for the heating. We need to get around the rest of the park and find the other teams.

As we enter the valley we hear bleeps from the radio.

'It's 962,' says Kira. 'He's a subordinate black adult. He seems to be travelling alone since losing his brother last year. We think he's around seven and a half – one of the older wolves in the park. His is the Eight Mile pack, a group of twelve, the ones that we heard howling. We think the younger ones are splintering off now. Perhaps they were trying to communicate with him, perhaps they know he is still around. It will be interesting to know what he ends up doing. Maybe he is just having some alone time or maybe he is leaving the pack too.'

I can't help but smile. Kira knows wolf biology, of course, but she is so familiar with the individual wolves and their family stories. Her observations over the years have also shown how their personalities affect their destiny, their choices and therefore the survival or not of a pack. To her and to me, this is so much of the fascination, so much of the continual intrigue.

'We are six or seven generations since the reintroduction now, and I know nothing compared to Rick – he knows all the lineages.'

She is talking about Rick McIntyre, a park warden and the human legend of the Lamar Valley. He has been watching wolves since 1996 and has barely taken a day off – in the first twenty years he missed two days for his mother's funeral and then had heart surgery and only missed five days.

'Hopefully we'll meet him today; he is usually out here.'

At Tower Junction, we stop the car as a huge group of elk leave the trees to our left and trot across the road. They are beautiful in the morning light, their breath steams, and the sun's low beams delineate their backs, highlighting the frosty coats they wear. They are prancing, as though the early-morning warmth is bringing them back to life and they are trying to defrost the ice on their backs. We are silent for a moment, just enjoying the sight.

When they are gone we continue, turning left back towards Lamar Valley.

'Only a handful of the wolves here have lived to be over ten years old,' Kira continues, 'and it really takes them close to two years to learn what they need to learn from the other members of the pack.'

Our chat is interrupted by the radio. Kira pays close attention to the regular contact as the teams check in, informing us of group names, numbers and whereabouts, while she also listens out for collar transmission bleeps for wolves. We hear: 'Wapiti at Mount Hedges.'

'Ha!' Kira smiles. 'The Wapiti pack are the biggest pack right now. We think they are seventeen strong. Once they get above fourteen or fifteen, a group – often of same-sex siblings – will split off after about six months and start their own pack. Some really have wanderlust and go super-far. They have been known to get as far as South Dakota, Denver, even the Grand Canyon. That behaviour has to allow for genetic diversity, so it's good. It usually works that the girls stay in the pack, but not always.'

Kira could obviously talk about wolves forever, and I could listen forever. For every wolf rule, there seem to be exceptions, and the more I hear, the more it seems there is to learn.

We pull over at Slough-pronounced-Slew and Kira says, 'The Junction Butte pack are here, I think. This is their den.'

There is a group of cars. It may be early but the wolf watchers have already gathered; in fact, there is quite the crowd. Here also is Rick McIntyre. I recognise him immediately from everything I have ever read about him. He has red eyebrows, and I can't see his hair as it is under his hat, but I notice the thread veins in his cheeks from being in this harsh climate almost every single day of his life.

Rick and Kira exchange warm greetings – they see each other most days – and I am introduced. Before I know it, he has taken my scope and is adjusting it. He apparently retired last April, but that has made no difference at all to his schedule, which remains the same. It will probably always remain the same. He looks up from my scope and smiles.

'The wolves are doing nothing,' he announces.

He is exceptionally softly spoken, with a kind, open face. There is a gentleness about him as he peers into the distance towards the pack. His eyes are the same blue as the clear winter sky.

'The pups are pretty much fully grown now,' he explains, 'but this is where they were born, and they all turned up this morning to check out their old den.'

The Junction Butte pack are now resting in the tall tan grass on a slope a few hundred yards away. Looking through my scope, I don't seem to be able to see them, just a few slabs of rock and lots of grass poking out from the snow. Then I realise that there is actually only one slab of rock – it has taken a moment to 'get my eye in', which isn't helped by the fact that both eyes are watering because of the cold – but the rest of the rocks are actually wolves just lying around.

As we watch, a couple of the pups, restless, get up and start nosing around.

'That's right where the den is,' murmurs Rick.

They disappear for a second, obviously down into the den. I wonder what they are thinking – does this feel like coming 'home' to them, or are they just curious? It seems to me that, given the complexity of their relationships and the long-term committed nature of wolf family life, they might well be attached to this place where they spent their 'puphood', and that they probably do feel more than curiosity for it. But that is what is so fascinating about it all – however much we want to, however much we are drawn to them, we can't quite get inside their heads. Still, the work that Kira and the rest of the study team are doing with census and observation brings us closer and closer to understanding.

Clearly, after yesterday's excitement and all that long-distance travel, the Junction Butte pack are having a pyjama day: busy doing nothing. Another piece of the jigsaw comes together.

We pack up to go and find the next group of biologists, with a brief pause to jump start a fellow wolf watcher's car before we leave the car park. Waved off by some grateful folk, we are soon off again into the Lamar Valley.

Keeping our eyes peeled for wildlife, Kira and I discuss the way wolves hunt. She explains that a 2009 study of the Yellowstone wolves showed that they have different roles in a hunt, depending on their size and age. The larger wolves – usually the males – tend to have the advantage when it comes to grappling with and subduing prey like elk, but that size can be a disadvantage when it comes to speed. It seems obvious, I guess, but there is a point when bigger is not better: it's not useful being big enough to take down a big elk if you are too big to outrun it in the first place. So a combination of sizes in the pack means different wolves take different roles; some are better at speed and chasing, others at taking prey down. Kira mentions, too, that the younger wolves are also generally the fastest.

We talk about 926, daughter of 06. I tell Kira how she was spotted in Silver Gate and how Keith the chef was so delighted to have taken such close-up photos of her. Kira sighs.

'It can be a problem. We are worried about the Lamar Canyon pack because they are so comfortable around people. In the park they are safe to walk right down the middle of the road, but not in Silver Gate. There is talk of hazing them, perhaps to avoid that, to change the behaviour and keep them safe outside the park. 06 was comfortable around people too – she had figured out perfectly how to live in Yellowstone, but that didn't translate when she went out of the park, and of course she was shot.'

I ask what happened to the rest of her family.

'Well, 06's mate's life as a widower never really worked out. He lost one mate to a hunter, then to other wolves, and then one beautiful white wife ran off with three others and started the Wapiti pack.'

Our time together continues in this way – listening, talking to watchers and biologists alike, fitting together the pieces of the complex puzzle that makes up the lives and interactions of the individuals and packs here. I am catching up fast with pack names, wolf numbers and their relationships. I am more intrigued every minute.

CHAPTER FOURTEEN

The wolf watcher's surprise

On my last day the weather has turned. The snow is coming in, clouds and visibility are low, and we are forecast a winter storm. Given that there won't be much wolf watching for the rest of the week, I have decided I am keen to get back to Jackson for Thanksgiving the next day.

I need to get that five-hour drive done before the snow gets too bad. I will have to get over the Teton Pass to get home, which is very often closed if the conditions are too bad. I travel through the park alone and barely see another car. However, I am most of the way along the Grand Loop Road, which takes me back to Mammoth and the north entrance, when I spot some familiar cars in the Blacktail Creek Trailhead turn-out. I pull in to see what is going on.

There are no scopes out, just a gathering of familiar faces. Rick is inside his car and has a scope attached to the car window but isn't paying it any attention. Instead, he is speaking softly into a radio and taking notes. I nod to say hi but don't disturb him. He is in classic Rick wolf-research mode.

Rick Lamplugh is there, author of beautiful books about Yellowstone and wolves. My favourite is *In the Temple of Wolves*, which describes his time in the Lamar Valley, and I am thrilled to meet him. We joke about 'the addiction'. No one seems to get away without being bitten by the bug. You can write as many books as you like on the subject, but you will still never be done with coming back to find out more about wolves.

An older man pulls up. He is such a star in the wolf-watching world that he has been given the nickname Wolf Man Cliff, and is such an addict that he actually moved to

Cooke City. He is now seventy-eight and tells us stories of how, when he was younger, he would trek all day all over the hills and mountains of Yellowstone on the trail of the wolves and other wildlife. I can see his frustration at his age, as he tuts and says, 'I wish I could cover the ground as fast as they do.' Wolves can easily travel twelve miles in a day and, as I am beginning to discover, they don't tend to just stay in one place.

While we set up our scopes, we catch up on the wolf news. Cliff says that he heard the Wapiti pack were up here somewhere but wasn't sure if they could be seen. Judging by the way Rick McIntyre was ignoring his scope, that was probably the case.

Then we fall silent.

Up on Blacktail Plateau a wolf is howling. It comes from the direction Rick McIntyre's scope is pointing towards.

We wait, silently adjusting our scopes. I almost hold my breath in case the noise of it might prevent me from hearing an answer or joining chorus, but there is nothing. There is nothing to be seen through the scopes either, nothing but rolling land with lots of undulations that could easily hide a distant wolf.

The wolf calls again.

No reply.

'Who could it be, any idea?'

'Maybe one of the Wapiti pack. They were here,' says Cliff.

Rick McIntyre joins us briefly.

'No,' he says, 'we just had news that Wapiti are down near Tower Junction now.'

A few minutes later the wolf calls one more time, and that is it. After that, there is silence.

The others pack their scopes, ready to head back into the park to see the Wapiti pack. I am so tempted to go with them, but I didn't even schedule this stop, and I still have a long drive ahead of me in a winter storm, so I have to leave on a mystery. I guess that is how the addiction starts – just one more wolf, just one more answer to a question.

The following evening, full of Thanksgiving dinner, I arrive home alone. I can't help going outside to the deck to look at the moon and am suddenly very conscious that I am far, far away from the Lamar Valley.

I tell myself that it's unlikely, but there could be the odd wolf out there right now. Part of the joy of living here in Wyoming is that you never truly know which animals are out there in the darkness. However, there is no certainty about wolves here – not like being in Silver Gate or the Lamar Valley – and so it is the absence of them that I feel. If I could hear them howling, I know it would give me comfort. There is some solace in knowing that there is another tribe out there experiencing life in a similar way, feeling the bonds of family and friendship, dealing with the unexpected challenges that life throws at them.

I yearn for something. But what? Wolves, the wilderness? Something other, bigger, mysterious.

Perhaps this is why people turn to God. Maybe if I believed in God, I might feel OK again – knowing that something was out there in the darkness might fill the gap that loneliness has made.

Instead, the dog, who stays by my side whatever I feel, presses himself against me. He has a million places he could be on this deck and in this night and yet he chooses to be less than a millimetre away.

'Not god, dog,' he seems to say.

What is that? What do we share with canines? Does he really understand? How about the wolves – do they feel things the same way we do? They have family bonds for sure, and they rely on them for survival. But do they feel it like us?

My 'loneliness', this yearning I feel, is appeased in the Lamar Valley by just knowing that they are out there, just getting a small glimpse into their lives, their families. Wolf society relies on teamwork, family, loyalty, unconditional love – the things I am missing right now.

Those things are tangible in wolves. Others may fill the gap with religion, but I'm not sure I can. I suspect there is something bigger than us – I hope there is – but I can't see it, and it isn't filling me up right now. Wolves I can see. I can watch their pups at play, teasing the adults till they give up on rest and join in. I understand that. I have been the adult settling to digest after a big meal. I have been the child who isn't ready to sit still and wants more fun and attention. I remember watching my Auntie Diane dozing by the fire and wondering how on earth she could keep her eyes shut for so long in the middle of the day.

Back inside, with the fire lit and a glass of wine, I let Doodle rest his head on my hand, stopping the pen from moving over the page, looking deep into my eyes as he tries to disturb me.

'What is going on with you, Mum?'

We have a link. We have a mutual fascination. We want to know about each other. We suspect we are alike, but we just can't talk it through. Something other… but something real, relatable and tangible. More than ever, I want to understand wolves and to understand our fascination with them. Society and family is everything to us humans. Our place in it, the love we get from it. It appears to be the same for wolves – even a lone wolf is always looking to join a pack.

I have my dog. We have a mutual understanding – well, unless he has found an old elk bone and doesn't want to come back to the car. He needs me to feed him and keep him safe, to have fun with him. Yet the wild wolf has no need of us, he is self-sufficient. To really understand the wolf might perhaps be to know that we – as a thinking, caring species – aren't alone on this planet. I'm sure elephants or whales could give us the same, as they are intelligent and social mammals too. But we don't have a semi-elephant living in the house with us – we haven't become that close to those species.

We are not all, as humans, so self-assured that we feel complete. It depends on our story. And so I believe it must be for the wolves. After this week of wolf watching, am I taking it too far to think that they seem to have the same drives as us – to be needed, loved and accepted? Those things must drive each individual to want to be part of a pack, and that is how they survive. We are the same, we can't do it alone. For my family to really work, I need my alpha.

Weirdly, I can't stop these thoughts of wolves and family spinning around my head, even as I climb back into my own bed and snuggle down to sleep.

The next morning I wake to the surprise news that 926 – Spitfire, daughter of the legendary 06, leader and mother of the Lamar Canyon pack – is dead. She was shot in Silver Gate by the man who owns the Exxon garage next to Soda Butte Lodge, right where we parked. He 'legally harvested' her, and she is now his trophy.

In her absence people mourn, wolves howl and a family – a tribe, a pack, however you want to describe it – is in disarray. For a wolf pack to work, they need their alpha. I wonder if the hunter even thought of the cost to the family of wolves. I wonder if he knew or cared. I wonder if I am being anthropomorphic.

Only a few days before 926's death, Karen I had been thrilled to walk in the footsteps of this strong, female leader of her wolf family. The sadness stays with me.

CHAPTER FIFTEEN

Hunting the hunter

It is spring as I head through Yellowstone toward Silver
Gate again. The wolves there are on my mind, and I can't
resist the urge to catch up. It is as though I have unfinished
business, somehow.

This time as I drive through the park, I am slowed not
by snow but by bison. They are on the move, everywhere.
One tribe even acts as though they are part of the traffic,
walking down the right side of the road in single file, taking
great care not to cross the solid yellow line in the middle
that means no overtaking, as a line of cars builds up behind
them. One herd leads into another as if a great bison
gathering is taking place, and I see lots of babies – cautious,
skittish, but staying close to their mothers.

On either side of the road, patchy snow is still on the
ground, but it is melting fast, revealing the dead brown of
the grass that died at the end of the last season but will soon
turn bright green in the longer days. The road itself is clear
and dry, the sky blue and the sun warm.

Cooke City always has a surprise in store. Today, the
first tiny log cabin I drive by in the 'high street' has a
large white notice in the window: 'Tuxedo and Ballgown
rental opening soon'. Really? Here? What for? I wonder if
the Range Rider is opening its doors for a new kind of
business.

I'm nervous about visiting Cooke City this time, as I
have a different agenda. Last time I was excited to go wolf
watching, staying in the cabin with Karen, and keen to find
out more about the canine inhabitants of Lamar Valley.
This time I want to find people, I want to ask questions –
questions they might not want to answer. The trouble with

wolves, I am fast discovering, is actually the trouble with people. They are really my trail.

I have been reading as many articles as I can find about the wolf 'situation' in Silver Gate and Cooke City ever since the death of 926F, and a few names stand out. I know, for example, the name of the man that killed her. I know where to find him, and I want to ask him why he did it and whether he has any regrets. I know that since then there have been many discussions about whether a Yellowstone wolf should be permitted to be shot just over the park boundary in Silver Gate, not only within this small community but in the wider community outside of it. Locals are suggesting that perhaps there should be a 'buffer zone'.

Cooke City and Silver Gate, sitting right next to Yellowstone as they do, rely on tourism, and most of those tourists are visiting the park to experience the wildlife. Given the fact that the Lamar Valley is just a few miles away, it's not hard to work out that most tourists are hoping to see wolves. Social media has not been very forgiving – there have been calls to boycott the 'wolf haters' of Cooke City, and this whole scene has frankly been pretty awful PR for the place.

I'd like to hear what everyone has to say, and most of all I'd like to listen to what the hunter has to say. I've put this trip off for a while, though. I couldn't quite work out why, until I realised I was scared. I was scared he might not speak to me. I was worried about confrontation, about putting more pressure on people. I didn't want to be a snooping journalist, but I really wanted to understand the truth.

'So what's the worst that can happen?' I ask myself as I stop to turn left at Tower Junction and wait for a herd of bison crossing the road. 'The worst that can happen is that he won't speak to you or that you find no one who will speak to you. But then at least you know you tried.'

I have to sit with this thought for a while as the herd is pretty big. The tiny-eyed bison plod by, wide-eyed calves

obediently following their mothers, before they all settle to
graze on the other side of the road.

I drive past Slough Creek, where there are a few cars
parked up, including Rick McIntyre's with the large
radio aerial sticking up from his roof. The Junction Butte
pack must be hanging out at their den again. I realise it's
that time of year already – females must be pregnant and
thinking about denning. Tempting as it is to get the Junction
Butte gossip and lay my eyes on that healthy pack, I drive
on by. If I stop, I could be distracted for the rest of the
afternoon.

If the Junction Butte pack weren't doing so well, perhaps
the Lamar Canyon pack would have spent more time in the
safety of the park and not been driven east and got themselves
into this mess.

'I will drive straight through Lamar,' I think. 'I will not
stop. I will face the uncomfortableness of trying to get people
to talk.'

Only, of course, I don't. Instead, I see a grizzly.

He is way off in the distance. At least, I presume it is a he
because, as much as I peer, I see no signs of cubs. From
this far away, there is no other way to tell. He is in the trees
on the other side of the creek, head down but not foraging,
just making his way through the snow and heading east –
the same direction I am going. Every now and then I lose
sight of him between the trees, but then I see him again.
He knows where he is headed and he isn't stopping.

After a while I sigh and start the car again. For the last
few months, I have watched people hurling insults at each
other from behind the safe walls of Facebook, but now
I want to have a face-to-face – it is different.

I drive through Silver Gate and on to Cooke City, where
I pull up outside a log cabin cafe called Buns n Beds. I know
who I am hoping to find and this is probably the best place
to start.

'Not my job to judge,' I remind myself, 'just to listen. Just to discover.'

For a second I miss my dog.

I step out of the car and into water. Under the new blue sky, the snow is melting fast and the 'high street' is part-stream today. I take a deep breath and walk into the cafe.

Only two people sit at one table. They look up as I walk in.

'Hi, welcome!' they say in unison in that typically American fashion. Only after living here have I realised how sincere it is.

'Hi. Might you be Jan?' I say.

'Yes!' says Jan and shakes my hand. 'What can I do for you?'

I explain. The man she was talking to leaves immediately.

'Would you like some lunch?'

'Sure,' I say, glancing at the menu. 'I'll take a cheddar bomb burger and chips. And a beer.'

Chips are always my go-to when I am stressed.

I am relieved to discover that Jan isn't the least bit fazed when she hears about my work. We make small talk while she cooks and I sit at the counter with my beer and notebook.

The cafe is cosy and full of signs like 'We don't serve women, bring your own' and photographs of local 'celebrity' wolves and bears. There is even a bear chart that documents the local bear families. One cross-country ski stands by the door for knocking the snow off the porch.

'I'm not putting it away just yet,' Jan says with a wry glance. She and I both know that a little bit of spring doesn't necessarily mean we are done with snow.

The tablecloths are perfectly clean plastic with green and white stripes. Outside, beyond the neon 'open' letters, I see signs for the Miner's Saloon and Beartooth Ally, the American flag wafting in front of the 'Bistro', and another

cabin-type building. Before I know it, lunch is served in a
red plastic basket, and I remember at the last minute that
'chips' means 'crisps', and 'fries' means 'chips' – doh! I'm not
usually a cheeseburger kind of girl, but Jan has her own
homemade sauce and it is fantastic – a mix of pickles and
maybe dill and something else I can't quite identify. The
meat is local, soft and succulent, and the flavour is so unique
that I know I'm going to crave one of these when I'm
hundreds of miles away.

'We haven't really spoken about it face to face,' says Jan
as she wipes her hands and I take another giant mouthful of
cheese bomb burger. She is referring to the hunter, her
neighbour. 'We are all getting a bit shy of the publicity, to
be honest, wherever we stand on it. We are such a small
community. The thing is, I am a dog owner. I love dogs, and
so I will never understand how anyone who loves dogs can
shoot a wolf. I struggle to think of shooting any animal.
Even if I had to do it for meat, I'd go veggie! I was a vet
assistant – I wasn't always a cafe owner – and in those days
it was, "you shoot 'em, we save 'em," I guess.' There is a
pause while she thinks and I chew. 'Look, he is a trophy
hunter, which is something I won't ever understand, but she
wasn't even that pretty a wolf. Yes, she was well known and
loved, but not actually that pretty.'

Outside, a quad bike comes up the middle of the road
with a white dog sitting happily on the back, ears and
cheeks streaming behind him, mouth smiling. The door to
the cafe opens and in walks a customer. I know better than
to keep talking – Jan has work to do, and our conversation
is private.

The new customer is Debbie. She just moved here two
weeks ago after deciding to restart her life at sixty-one. She
is grey-haired but energetic and excited and is busy making
friends. As she chats, telling us all about her pets and her
long drive hampered by snowstorms across many states,

I notice her sweatshirt has a picture of a wolf on one side and the American flag on the other.

Jan is busy creating another cheese bomb burger.

'Hey, I just noticed the wolf on your sweatshirt,' I say.

'Yes,' she sighs. 'In honour of my friend Nathanial.'

'What happened to Nathanial?'

'Well, he and two friends were camping up in Mackenzie National Park in Canada many years ago, and they found three wolf cubs. They figured they were orphans, so they took them home with them. I mean, this was before it was illegal to do that sort of thing,' she adds in response to my raised eyebrows. 'So they each took one and bought 'em home and raised 'em. His was called Natasha, and it isn't true what they say about 'em – they really make the most amazing pets. She was perfect, his Natasha. A big white wolf, part of the family. I mean, in every family photo she was there right at the front.'

'So, what happened?' I ask. The smell of burger wafts across the room and I glance at Jan. Although she is busy behind the counter, I can see she is also curious.

'Well, because she was a wild animal, she couldn't have a rabies shot, so the local cop shot her.' She takes a sip of her drink and shakes her head. 'It was so sad. I love the wolves. I moved out west just because of all the wildlife.'

I gaze out of the window and wonder about this. I notice the fuel station has a sign next to the pumps advertising buffalo, elk and venison jerky on sale inside – not something you would see at home. Nor is the Gator 4x4 at the pump.

I know that one of the reasons wolves were hated in the early days was because they could spread rabies. I remember reading stories of men who had been bitten and gone crazy. In those days they didn't know what rabies was – they just blamed the wolves.

Another burger served, Jan wipes her hands as she sits down beside me to talk. She also lives here because of the wildlife, she tells us, and describes the foxes that den in her garden, the joy at springtime of watching the fuzzy grey cubs playing just outside the window while she enjoys an early-morning coffee.

'You know,' she says, 'we didn't have foxes before the wolves came. At least, we didn't see many; the coyotes kept their numbers down. But now that the wolves are here the coyote numbers are down, and so the foxes have a little breathing space to flourish. Before the reintroduction, no one really understood all the differences that the wolves made to this place.'

We discuss this a little. Yellowstone, Montana and Wyoming had huge elk herds and many more moose before the arrival of the wolves in 1995. Most people would have predicted those herd sizes being negatively impacted by the appearance of a predator – that made common sense – but no one really understood the extent of the positive effects that were seen throughout the ecosystem. Elk numbers came down, but their behaviour changed too. Now that they had to keep an eye out for predators, they behaved in a more natural way, travelling in herds that were smaller and tighter rather than spreading themselves across the landscape. They began avoiding the places that left them vulnerable to a pack in certain seasons. The constant pressure from the presence of predators kept them on the move, so they stopped staying in one place and intensely grazing it until the food there was done.

What we are discussing has a name in ecology: the 'trophic cascade' effect. Imagine the ecosystem as a triangle with the wolves at the top. The wolves' presence or absence changes the interactions and survival rates of all the species below them in the triangle, right down to invertebrates and plants at the bottom. Willows now have the chance to rejuvenate and flourish rather than being eaten as they grow. That, in turn, has improved things for many other species. The

abundant willows are food for beavers, so they are returning
to the park. Before the wolf reintroduction, there was
only one beaver colony; now there are nine. The return of
beavers and their famous dams changes the structure and
function of the waterways, creating cool shady pools for
fish and evening out the seasonal snow run-off. Songbirds
have come back, too, now that there is willow habitat for
them to live in again. Regular wolf kills on the landscape
provide food for scavengers like ravens, magpies, eagles and
bears. Before the wolves came back, those creatures were
relying on a boom-or-bust situation because elk or bison
only died when the weather conditions were too adverse
for them to survive.

As time passes, we are learning more and more about
the positive impacts and the extent of the 'wolf effect' on
the whole of the natural system. Jan is no ecologist, but
she is local and loves wildlife, and that gives her all the
credentials she needs to notice the differences. She understands that wildlife has to be managed and monitored, and
she respects Montana state for doing a great job with the
challenges of that.

Debbie finishes her cheese bomb, thanks Jan and leaves
with a cheery goodbye. We are alone again. I have been
waiting for this opportunity to ask her what she knows
about the dog that was taken in December.

'Well,' she says with a sigh, 'it was really sad. Just a few
weeks after 926 was shot, one of our neighbours here got
up at around two in the morning to let her dog out for a
pee. He was pretty old – around fifteen, I think, and very
arthritic. Anyhow, the wolf pups attacked him – I guess
they thought he might be easy prey. When my neighbour
heard the commotion, she shouted and yelled and the
wolves left, so she managed to get hold of him and take him
to the emergency vet, but his injuries were too bad, and he
had to be euthanised.'

'That sounds awful.' I can't imagine how it must be to witness such a sad demise of your dog, how horrific it must be to hear the howling and screaming of such an attack.

'It was horrible, and we all rallied around. Everyone has so much sympathy for the owner.'

'Why do you think they did it? It was out of the ordinary for them, right?'

'Yes, but you see the problem was that they were tracking their mum. They had been howling for weeks. They were looking for her in all the places she had been. You know that back road in Silver Gate?' I nod. I know it – that was where Karen and I had followed wolf tracks. 'Well, right up there just past that tennis court where that really big house is, that is where she was shot, so they were tracking her all over. I think they tracked her that night. They even hung out there for a while, in the place that she had been shot, just howling for her. I think that is what they were doing when they came across the dog. Perhaps if their mum had been there, she would have warned them off, and they wouldn't have bothered, but they are lost without her. She was the leader of their family. Now things are different, and they don't really know what they are doing.'

The wooden door opens to a familiar face.

'I heard you might serve food around here? It sure smells that way.'

It is Wolf Man Cliff, one of the seasoned wolf watchers I met in a turn-out in November. He remembers me, greets both Jan and me warmly and sits stiffly down at our table. Jan brings him a Coke.

'Don't forget,' he says to me, 'I'm a foreigner. I can't speak English like you do!'

He had spent the morning in Yellowstone in the sunshine and caught up with the infamous Rick McIntyre in the Slough turn-out, so it is no surprise that our small talk turns

swiftly to wolf talk. I am keen to get the latest news on the
struggling Lamar Canyon pack.

'Have you seen them?' I ask. 'How are they faring?'

'Well,' he replies in a deep drawl of a voice, 'they were
five pups, and now it seems they are down to two. They
have a female with them, and she is Little T, who seems like
she might be looking for a den and maybe going to have
some pups, but Small Dot hasn't been seen for the last
month or so and he is the male so I don't know what has
happened to him. The four of them were singing for a
while here in Silver Gate, but we haven't heard them lately,
so we aren't sure where they are.'

I wonder what that means for the future of the pack.
How can one female have pups and still look after the
yearlings? I know that they don't have enough experience
to look after her yet. It could be that the future of the pack
depends upon that female not having pups this year or it
could be the beginning of the end for them, and they will
disperse. Perhaps this process has already started if there are
three pups and Small Dot already missing.

'OK, how are the Junction Butte pack getting along?'

Their strength appeared to be the Lamar Canyon pack's
weakness.

'That pack appears to be denning at Slough creek again,'
says Cliff, stroking the white stubble on his tanned, crinkled
face. 'They may already have pups in that den. In fact,
I think it might be 907 who has had them. She is a pretty
dark female, if I recall it right. Anyhow, they haven't stopped
travelling between their favourite spots. They were at Hellroaring this morning and on a bison kill a couple of days
ago up in that backcountry. You couldn't see it from the road,
but I knew where I could get to on the opposite hill without
disturbing them, so I got a good view of them there.'

So the Junction Butte pack are still claiming the large
territory they were using when I was here in the winter, still

visiting the same spots regularly despite the other packs, and still covering a lot of ground to do so.

Cliff takes a slurp of Coke. 'Do you know the children's fire trail?'

I shake my head. There is still so much I have to learn about Yellowstone.

'That's a trail up near the north, further towards Mammoth. In the summer you can get right round there, and that's where the Eight Mile pack have been hanging out. I've seen them lots in that area and south of there. I think nine of them were up there looking for a den site about a month ago. There is a black bear den near there, and about four miles away from there, the Wapiti pack just showed up.'

We sit quietly for a while, Cliff's kind rheumy eyes kind of distant.

'That wolf that was shot was no threat,' he says eventually. 'The balance seems to have gone out of it all somehow. It's just too extreme on both sides. Bears are more dangerous than the wolves. I know that management is important. This is a rancher's environment, too, and these guys,' he nods at Jan, 'were all for creating a buffer zone in Silver Gate and Cooke City, but the state turned it down. And I understand it to some degree: if they give them one buffer zone, before long they'll just want another one, so where do you stop? I'm not one of those environmentalists who is into too much government intrusion, like some of the wolf watchers. And some of the tourists can make it all very difficult, so there's controversy now over them getting too close, and rules that should be in place and that is interfering with our constitutional rights. People who don't know about wildlife get too close, but I'm not one of them, and I don't go with over-management to the point where we can't go hiking where we want to in the park.

'Then there is so much ridiculousness on the internet.' He practically spits the word out as he gets into his stride.

'So much bad-mouthing and nasty stuff on Facebook and social media, accusing the wolves of stopping the traffic on the road and saying that they should block the road if the wolves are there. Sometimes the advocates act like a big clique; sometimes the others do. I'm sure that the rules of wolf watching are going to change now that McIntyre is retiring. Already people – especially those photographers – just get too close.'

Jan chimes in: 'So what is the balance? We have to keep the animals wild; otherwise, we end up in this situation where they are too accustomed to being close to humans in the park, and then when they come out they aren't protected. One idea is that we haze them – both wolves and bears that are getting too close.'

I know about hazing; it happens in town. If a moose is too close to the school, they will follow it, making loud horn noises to scare it away, almost a pressurised form of herding.

'There are just too many people now,' says Cliff. 'It's all going to lead to a lot of restrictions. There were over two hundred people on the radios for the wolf watchers the other day. I'm a loner, I don't like too many people. I like to get out there into the wilderness. One day at the Lamar River confluence – this was back in the nineties when you could see the Druid pack every single day – there was just a ton of people parked up there. I kept away from them. I was on the hillside right above my vehicle, and I could see the Druids were across from me. Twenty-one of them crossed the road and went to the denning area. Anyhow, I got to know this fella with the same camera as me, a first-time wolf watcher. He had evidently seen me up on the hill and asked everyone "Who is that?" Do you know what they said?' I shake my head, smiling at his irritation. 'They said, "You don't want to talk to him, he's an old grouch. He doesn't like anybody." People will believe that nonsense. I just like keeping my distance because I want to watch the wolves peacefully. They all make too much noise, what with the radios and chat and photography, explaining it all. They are making all the noise and distracting me, and they will ruin any sound recording you are trying to make.

'I remember another day,' he says, on a roll now as he thinks back over his years of wolf watching. 'This was back in '95 when I was watching the Rose Creeks. They were the first ones out of the pen, and I started seeing them right at the beginning. Bob Landers was there too, do you know him? He is a film-maker, and he was taking shots with me.' I nod my head. Bob Landers has been filming wildlife in this area for years. 'There was a carcass right at the confluence across the creek, and in those days there wasn't really anybody there but us. I had never seen a wolf in my entire life. I didn't need to know what I was doing, though. It's not difficult to learn the habits of those wolves – I just learn from going and doing it. I never ever get tired of seeing 'em. I just like the outdoors and searching for and seeing wildlife.

'I used to walk out at Slough Creek to see the Rose Creek pack when they were there, and they were a big pack, around twenty or thirty wolves. I'd hike down the road or go west to the confluence of Soda Butte and Lamar. This day there was a carcass, and I hung out in the sagebrush on the south side of the river. The carcass was there, and the wolves were on the hill on that ridge. So I stayed four hundred yards from the carcass and hung out near it, but I was on the opposite side of the creek so I knew I wouldn't interfere with the wolves. It was cold, and there was snow on the ground, but I knew that they were going to come down eventually. In the end, it was noon when one came down and had a sniff at it then went back, and then at four they all got on the move, and a string of over twenty of them came down the ridge and started milling around the carcass. It was a fantastic thing to watch, and they didn't even know I was there. But then they left the carcass and came to the creek and crossed right in front of me. Right in front of where I was hiding. It was the first time I had been that close, not even seventy yards away. By then it was starting to get dark and my camera had run out. Jeez, it was cool, but I was getting really cold, and I knew I couldn't stay there in the dark, I had to get back to my car. I wasn't nervous – I just stood up so that they could see me. They went back and watched me while I left. That's what I like, I liked it when it was just me.'

Cliff smiles broadly for the first time, and his whole face lights up. I realise what a handsome man he must have been in those days. The cafe is warm and cosy and smells of good food, and I have no place to be and nothing else to do but listen to his stories. With a smile, Jan brings me another beer.

'I don't like a lot of research. It's too overdone, it seems like they are putting collars on everything, and I'm worried that will lead to more controls and limitation, and then there's going to be too much protection and rule-setting and shutting of areas and people's rights of access. It's fair enough if there's a den – I mean, that needs protecting – but how much do they need to learn? I just can't understand why there is still so much doggone research.' He laughs again. 'I learned what I have learned by just going out there and observing, by being gentle and quiet and just being there with them.'

In my mind's eye, I see Cliff out there, all those hours, all those years spent just becoming part of the landscape. I think of Kira and the other scientists at work and all that we are still learning from the research – more than a decade of information about the responses of an ecosystem to the absence and presence of predators. But I see it from Cliff's view too – should we interfere as much?

'See, at first we had a thirty- to forty-yard distance we had to keep from wild animals, but now that is up to a hundred yards. The place is getting overrun with people now, mainly here to see the wolves and the bears, but this is the public's land, not a business operation.'

'Most people are here for the wolves and bears,' agrees Jan, who has been nursing a coffee while she listens. 'And I feel like that is part of the service we provide in here and in Cooke City generally. People travel from a distance and come in here to eat, or they come in after they've been in the park, and the first thing they want to talk about is the bears or wolves or bison they have just seen. So I pass that knowledge on, and those people who have come such a long way are so grateful when they get to see something special, they will come back in and thank me, and I get to keep track of the wildlife and talk about it while they are here having a good meal.'

Tourism is big business for Cooke City; the residents rely on it. Some of their income is from the snowmobilers – the nearby Beartooth Mountains offer some of the best snowmobiling country there is – but an awful lot of it is from people who come to see the wildlife, so Cooke City and Silver Gate stand to lose a lot if they gain a reputation as a place of hunters and wolf haters. And of course, most people who make their home there do so because they love the wildlife. The hunter has risked his community in that way. I really want to ask him why.

'Have you had a favourite wolf?' I ask Cliff.

'I like all of them, I love all wildlife. But wolf 21 was really cool. She was so well known. I have tons of footage and pictures of her. And I do like certain packs like the Wapiti pack and the Mollies. You know, when the wolves are a good size and in healthy condition and they are a big pack – that's what I love to see.'

Cliff is tired, and so we say our goodbyes and hope that we will meet again on the road in Yellowstone. Jan is about to close up her cafe, and I have other fish to fry.

'By the way,' I remember as I am about to walk out the door, 'what's with the tuxedo rental shop that's opening? Is there any need for that here?'

'Oh, that's just a joke,' she laughs. 'You know what we're like here.'

I know where the hunter works, so I head over there. The only story I have been told is that he had a call from a local person who must have seen 926, and he was seen heading down to Silver Gate almost immediately afterwards on a snowmobile, then shortly after that, he shot her. She was an easy target, so it doesn't seem that there was even the thrill of the chase.

I am fascinated. I really want to find out why, but I am out of luck – he is still out guiding snowmobilers. The man who works with him is very friendly. I explain that I'm not a snooping journalist, that my book won't be published for a while yet and that all I want is to hear his story, his opinion, his motivation. He says they have had enough of publicity, that there have been death threats and that 'we hear what is going on in our own town, our own community, from the outside. There is too much drama. We just need to calm it all down.'

I leave my number and a short note. But the hunter never calls.

CHAPTER SIXTEEN

The wolf lady

'Well, if he won't speak to me, that's fair enough,' I think as I drive back through the park. 'I wouldn't speak to me if I'd had death threats.'

The sun is low in the sky, the park is beautiful. I stop for another herd of bison. Their usually dull coat is backlit, making them look a little fuzzy. This is a big herd, they just keep on crossing the road in single file, which is unusual – typically they are a bit more spread out. As I follow the line back, I realise they are all crossing a little wooden bridge over a creek, and they have to do it one by one. At last, only two bison remain, one is a small one. The small one has stopped – he clearly has no interest in going over the bridge. I wonder what is going through his mind. Has he ever even seen a bridge before? To be honest, looking at the rickety little bridge, he isn't the only one who is surprised by the fact that it has just supported a herd of bison.

The big one, now becoming impatient behind him, has horns. Perhaps it is his mum. Either way, the big one takes a few steps backwards and a few quick ones forward in a little charge, using those horns against the little one's backside to try to make her point. I giggle as he stands stiff-legged – he is not moving. After a few more goes at shoving the little one forward, the big bison clearly realises another strategy is required and so barges in front as if to say, 'look, I'll show you it is perfectly safe, just follow me,' and crosses the bridge. This has zero effect on the little one, who reverses, backing slowly away from the threat of the bridge and then – with a dance of decision – runs down the steep bank, through the creek and up the other side, crossing the road at a trot to catch up with his herd.

Finally. I can move on.

Until the next stop. This time it's traffic that slows my progress. I pull up, get out of the car to find out what they are all looking at, and realise that I am only twenty feet away from a black bear cub foraging for roots and food among the sagebrush. I crouch down to watch, not wanting to disturb him, leaning back against my car. The evening light lends a glow to his coat, bringing out the burnished tawny brown and lighting up the round ears. The bear has no care at all for the onlookers; it is as if there were no cars or people here at all. Respectfully, the humans keep a distance as we watch, all filled with the same delight at being so close to a bear. It is so quiet that I can hear the sound his large claws make as he scrapes at the ground, and his snuffling and chewing noises. He ambles his way slowly around the area, occasionally lifting his head and his long snout, but never acknowledges the presence of his audience. I take photos – it is impossible not to.

Eventually, he decides to move on and heads straight for me. By the time he is five feet away I realise I have to move – we are certainly not meant to be that close – but I don't want to startle him. Carefully, slowly, I walk around the bonnet of my car to the other side. Creeping round to the back, I wonder where he has gone for a moment, but then he pops out right beside me as if he has just got out of the passenger seat. Ignoring me, he saunters into the middle of the road where shocked humans also pull back to the banks and slip behind their cars. The bear just strolls right down the middle of the road without a care in the world. After around fifty yards, he turns right into the forest as if he were following an invisible footpath and disappears into the trees. The show is over.

A few hundred yards later, I pass the 'photographers'. These are not happy with an iPhone shot – they are here to take serious photos. They all look very frustrated as they point their long lenses into the darkness of the trees, where

Left: Spring on the banks of the Buffalo Fork River. I'm smiling with relief because, after a long snowshoe session, I have been able to offload the heavy kit for a while. Despite the smile, however, I am certainly still 'bear aware'. This was the first day we saw wolf tracks.

Below: I love living in and spending time in the mountains – you never know the gifts they will give you. This is one of my favourite views of the Grand Teton range turned to magic by the setting sun.

Above: One wonderful sunny evening in Grand Teton National Park, waiting and hoping that a family of wolves were napping in that landscape somewhere…

Left: On the Buffalo Fork River – this was the first otter we saw, a single chap, who certainly looked like he was getting enough to eat despite the weather.

Right: I love this photo of 'Wolfman Cliff' – his face lit up as he shared with me his stories of twenty-five years of watching wolves in Yellowstone National Park.

Below: Malou is an inspiration: a ranching mother who is breaking ground when it comes to living alongside predators. Here she is trying to persuade Annaliese to get into the truck while her daughters look on.

Above: We spent a long morning 'potholing' through snow that was often up to our thighs, up and down steep buttes, but when you are following wolf tracks, every step is worthwhile.

Left: When you know what to look for… Biologist Ken Mills pointed out these scratch marks on an Aspen tree. They were made by a bear climbing it.

Above and below: Early one November morning, this was the view of the Junction Butte pack through my scope, playing after a feast of bison. The following day Kira took the below photo from the plane. Led by five-year-old female 969F, the wolves were crossing Hellroaring Creek in northern Yellowstone National Park fearlessly heading into another pack's territory.

Above and right: At the Wolf Conservation Centre in New York state, Zephyr, the leader of the pack of the three ambassador wolves, is the first to come and greet me. After I have met them all, Zephyr takes his place on a high rock and is more than happy to demonstrate exactly what a wolf is famous for.

none of the precious golden light is to be found. I crane my neck to see what they are looking for and see another bear cub, but this one is way off in the distance and half-hidden by bushes. These must be the cubs that Cliff mentioned – the den site is maybe close to here. I haven't the heart to stop and tell the wildlife photographers what they just missed a few hundred yards away.

My thoughts, after swimming with images of beautiful bears peacefully foraging, quickly return to hunting. Had I a gun, and had I not been in the National Park, I would have been in the perfect position to shoot that bear and take it for a trophy. But there is nothing in me that wants to do that, and I know there never will be. The closest I can get to the pleasure that must bring is when I take a good photo. For me, that is the ultimate souvenir.

If I can't talk to the hunter who shot wolf 926F, then I will speak to others who kill for trophies. I need to find out why they do it. Conservation is so much more complicated than good guys versus bad. This may be the Wild West, but I'm hoping we have learned a lot since the day of cowboys and 'Indians'.

For the next week I base myself at Tom Miner Basin, a small community of houses and ranches on the north side of the park. The people who live here face similar issues to those bordering the Grand Teton park: wolves and bears know no boundaries and life has changed since they started coming over the hill. This is another place where the reality of wilderness hits human civilisation. It is also a part of Montana with a long tradition of hunting.

I'm staying in a small ranch house that is just like an English cottage. It has a large wood-burning stove as an antidote to the spring weather, which can still bring snow and wind here. Wolves and bears, cows and dogs roam the hills outside, and I feel at peace.

After many phone calls, it is here that I find the first hunter who will talk to me. I am invited to spend the morning chatting with Vern on his ranch. I'm not sure what to expect when I turn up. Will I have anything at all in common with a trophy hunter? Will he just think I am a tree hugger?

Vern is great. His generous face speaks of a lifetime outdoors, and his sparkly blue eyes welcome me into the home he built and now shares with his wife, a small collection of cats and a ton of stuffed animals. By which I do not mean a collection of cuddly toys – I mean real animals that are stuffed and on the wall, looking down at us as Vern makes coffee in his cowboy hat. I spot a dark grey and silver wolf pelt hanging by its teeth from the upstairs railing. This is perfect: a wolf hunter.

'I might not be the best person to talk to,' he says.

'Oh?'

'Well, I've kind of changed a lot. I used to trophy hunt, but now I'm all about soil health.'

'Oh, kind of poacher turned gamekeeper?'

'Eh?'

'No matter, I think it might be an English phrase.'

'Now I'm sixty I have a different thought process, I guess. I'm tired of killing. Killing stuff takes too much effort, and now I'm burned out on cattle and just train a few horses. The way I think now is that first, we have to have a planet we can live on and that everything comes from the soil.'

As he pours the coffee, Vern catches me gazing at the wolf hide. I want to touch it.

'Did you kill that?'

'Yes, I did. I shot it sixteen years ago.'

'So you are the wolf lady?' he says, setting a mug in front of me and settling in a comfy wooden chair.

'Ha!' I laugh. 'I wouldn't put it like that!'

I explain what I am trying to do, to find out why people are so passionate about this creature.

'What is it with this love–hate affair we have with wolves?' I ask him. 'I mean, most people haven't ever and will never get to see a wild wolf, so why do we care so much?'

'I don't see it is as a love affair with the wolf, but a love affair with a symbol of what we are losing.' I frown so he continues: 'We are craving not to lose some of the key players at the top of the apex.'

'That kind of makes sense to me. Go on.'

'The way I see it now is that we are all here to be stewards, good stewards.'

'I agree with that.' Somehow I already know that Vern and I are going to get along well. 'Yet we all come at it in different ways. For me, I look around at all the incredible wildlife and nature and think of the incredible design. And to me, if there is a design there has to be a creator; otherwise, if I'm here randomly, then I'm not accountable.'

'That makes sense.'

'So I don't choose to think that we are here randomly. I think we are here by design.'

'So do you believe in evolution?'

'I really question it. Some of Darwin's work I go along with, but not all of the rest of it. Have you heard about the law of irreducible complexity? It's a theory that states that evolution by tiny changes is not possible. OK, you could see it with Darwin on all his travels on the finches' beaks and all that, but how did the whole animal start? Surely you had to have a functioning structure in the first place for it to be modified? So where did that come from?'

Vern is a creationist. I have spent many years discussing evolutionary theory, but I want to hear him out. I want to respect his views even if I don't agree.

'Also, is there any proof that we weren't around at the time of the dinosaurs?'

Vern is smiling. He is baiting me, wondering how I will react. I raise my eyebrows.

'Look at this,' he says, reaching to a shelf and bringing down a dark blue New American Standard Bible. 'Do you know the Old Testament?'

'Not in detail, I confess.'

'Whoever wrote this stuff all that time ago – they hadn't travelled around the world, didn't have TV, had never found fossils and didn't know half of what we now know. So how did they write it? Do you know Job?'

'Not very well,' I grin. 'Bible studies have never really been my thing. I was more concerned with the origin of the species.'

We smile at each other – this is going to be fun.

'Well,' Vern continues, 'Darwin travelled the world, and he saw all these designs, but the idea he came up with doesn't really work. It doesn't make much sense to me. I want to show you something.'

I'm nothing if not open-minded, and I am definitely curious about why people think the things they do. So, within thirty minutes of meeting each other, Vern the ex-trophy hunter and I are studying the bible together.

This is a well-thumbed book, falling out of its hardcover. The pages are delicate and super-thin as Vern flicks through it. Suddenly it falls open, right at Job.

'This is exactly what I wanted to show you.'

I don't say 'it's a sign', which I would typically say to something like that, but we exchange glances, and his blue eyes are really twinkling now as he smiles. He begins to track a work-cracked finger down the page, which we both bend our heads over.

God is talking to Job. He is explaining that he is more powerful than anything on earth because he created everything. After telling Job to 'brace himself like a man', God then goes on to say that he is more powerful than the behemoth that he created at the same time as he created humans. The behemoth has bones that are 'tubes of bronze' and limbs like 'rods of iron'. God then goes on to describe the might of the leviathan, which no man could overthrow: 'If you lay a hand on it,' says God, 'You will remember the struggle and never do it again! Any hope of subduing it is false.'

'What if,' says Vern, 'we were around with the dinosaurs? What if, in our oral history, these stories were handed down? Otherwise, where did the man who wrote these things get them from?'

I read the passage with the Leviathan in it, taking in the detail.

12
I will not fail to speak of Leviathan's limbs,
  Its strength and its graceful form.
13
Who can strip off its outer coat?
  Who can penetrate its double coat of armor?
14
Who dares open the doors of its mouth,
  ringed about with fearsome teeth?
15
Its back has rows of shields
  tightly sealed together;
16
Each is so close to the next
  that no air can pass between.
17
They are joined fast to one another;
  they cling together and cannot be parted.
18
Its snorting throws out flashes of light;
  its eyes are like the rays of dawn.
19
Flames stream from its mouth;
  sparks of fire shoot out.
20
Smoke pours from its nostrils
  as from a boiling pot over burning reeds.
21
Its breath sets coals ablaze,
  and flames dart from its mouth.

22
Strength resides in its neck;
  dismay goes before it.
23
The folds of its flesh are tightly joined;
  they are firm and immovable.
24
Its chest is hard as rock,
  hard as a lower millstone.
25
When it rises up, the mighty are terrified;
  they retreat before its thrashing.
26
The sword that reaches it has no effect,
  nor does the spear or the dart or the javelin.
27
Iron it treats like straw
  and bronze like rotten wood.
28
Arrows do not make it flee;
  slingstones are like chaff to it.
29
A club seems to it but a piece of straw;
  it laughs at the rattling of the lance.
30
Its undersides are jagged potsherds,
  leaving a trail in the mud like a threshing sledge.
31
It makes the depths churn like a boiling caldron
  and stirs up the sea like a pot of ointment.
32
It leaves a glistening wake behind it;
  one would think the deep had white hair.
33
Nothing on earth is its equal—
  a creature without fear.

34
It looks down on all that are haughty;
  it is king over all that are proud.'

'If we were around at the time of the dinosaurs, wouldn't that be it?' says Vern.
  'Well, it sounds more like a dragon to me,' I reply. 'But I have been watching a lot of *Game of Thrones*.'
  We both laugh as he closes the bible.
  'All that I mean is that if we accept the possibility that we are here by design, then we accept the possibility that we are not here randomly, and so we are accountable.'
  'I see.' This is not what I was expecting to hear from a man who had hunted wolves. 'So how does hunting fit into that?'
  'Well, like I said, I don't care for killing stuff any more.'
  'How about when you shot that wolf? How did it fit with your thinking then?'
  'When you love wolves, you aren't really thinking about the whole system. When you hate wolves, you aren't really thinking about the whole system. When you live with wolves, you have to think about the whole system. There was quite a story to that wolf pelt.'
  I sit back in my chair. I had suspected as much.
  'I hunted it in Alaska. At that time, the Inuit were suffering from a lack of meat because of a predator pit. Do you know what that is?'
  'I don't, although I probably should.'
  'It's when a predator population keeps a prey population at a really low density. At this time in Alaska, there were a lot of wolves, and in the season I was there the moose were sinking into the snow – it was warm, and their feet just weren't working like snowshoes any more, and so they were like sitting ducks for the wolves – just really easy prey. The Inuit were, by law, no longer able to shoot wolves from a plane, and their caribou and moose populations were being

decimated by wolves, and they didn't have enough meat. So it was the sensible thing to do to bring wolf numbers down. It needed to happen.'

Vern glances over at the pelt before he continues.

'My dad was on that hunt with me. We tracked a pack of wolves for over two hundred miles on snowmobiles, and over that time we came to realise that there were twenty or thirty in the pack, it was big. Also, over the time we followed the wolves, we found seven dead moose along the way. Like I said, sitting ducks. The wolves had killed them and moved on. All we saw that had been taken was that the lips and faces were torn up and they had opened up the backside and eaten the ass. That could have been other predators or birds after the wolves had killed them, but the wolves weren't killing for food; otherwise, they would have hung out and eaten them.

'Then we came across a biologist caught in the ice. He had fallen in. His snow machine was under and he was trapped. We dragged him out. It was kind of a fluke we found him or he would be dead, but we lit a fire and got him into dry clothes and dried his clothes, and while we were sitting there we got to talking. He was actually checking snares and traps. He was culling wolves – the same as us, really – managing the numbers for the needs of the Inuit people, looking at the whole picture. You have to look at the whole picture, you see.

'At that time, I wasn't working for the Inuit people. I wanted a trophy. I wanted a wolf. I didn't think of it as a problem because there were too many of 'em anyway. I don't regret it, but I don't need another one. I just don't. I got my wolf, I don't need another.' We both look at the trophies around the room. 'Each one has a story,' he says. 'And like I say, I don't care about killing stuff any more.'

There are stuffed animals along the top of the kitchen cabinets. One is a lynx standing straight up, one back foot on a wooden stump, his front paw reaching high into the air to grab a stuffed grouse that is attempting to fly off. Next to it – climbing onto another log with a stuffed bird inside it – is a curious creature. It is as big as the lynx but more chunky, and dark brown with flashes of white on its shoulders. Its weaselly face looks too small for its body, with black eyes and nose and tiny round ears. Its head has been positioned by the taxidermist to look as if it is gazing down at us in the kitchen.

'Do you know what that is?'

'I think so.'

'It's a wolverine.'

'I've never actually seen one.' I take the time to study it – I can, it isn't going anywhere. I hadn't realised how large they are. They are pretty rare to see in the wild.

'That one fought with a mountain lion before I killed it.' We stare at it. Then Vern stands up to go and get something. 'I have a new passion now.'

He brings a photo album to the table and opens it up. It is full of pictures of cougars close up and all sorts of other wildlife, but then he turns a page, and I see a perfect portrait of a leopard.

'You didn't take that round here.'

We laugh.

'No! I did not. I took it in a jungle in Tanzania. It was so hard to take. It had taken us a long time to find that leopard, and when we did, we had just come in from the light to the darkness of the trees. I was worried about the depth of field, about the focus, about all of it. Was there going to be grass in the way? It all happened so fast that it was hard to know if I was getting everything right, you know?'

I nod. I know. I have lived with my husband for twenty years, and he is the most passionate wildlife photographer I know.

'I got buck fever. I almost couldn't look at the picture in case I had got it wrong.'

'What's buck fever?'

'It's when you get so excited that you start shaking. After you have been looking for an animal for so long and you finally get your chance at a shot... If you get buck fever, there is every chance you can screw it up. What happens if – after all this time, all this effort, with this one chance – you take the picture and it's not right? What happens if I screw it up? I've taken hunters out and we've got good shots, and I've seen buck fever get to them so bad that they will miss by a mile.'

'See, this is a real trophy to me,' I say. Vern nods.

We look at more photos. Vern is so enthusiastic as he shows me the other side of his work, teaching disabled children to learn to ride and to hunt elk for food. I have no problem with that. In fact, it makes me see that behind 'Vern the trophy hunter' is 'Vern the human being'.

A theme emerges as we look through the book. Children seeing mountain lions up close; children who would never ride ordinarily, up on one of his horses, watching elk; trail-cam pictures revealing sights that we could never see – mountain lions mainly. Vern is opening people's eyes to the wildlife around them. Am I too emotional? Too empathetic to believe that the hunter cares about wildlife, about animals? It's just that he thinks about them in a slightly different way.

'Do you want some soup? It's lunchtime. And shall I light a fire? It's chilly.'

I say yes to both offers, and we sit by the fire, eating deliciously hearty elk and bean soup. I'm struck by how homely the house is. Goldfinches visit the multiple feeders outside the windows, flashing bright yellow in the sunshine. Wide and fast, the river runs below. I find myself a little distracted by the tiny old calico cat who leaps into my armchair, purring and playing with my pen.

'I first found out about wolves when I was growing up in Pennsylvania,' Vern tells me. 'My dad worked in the dairy industry, and the dairy cows that died were taken for wolf feed.'

'Eh?' I say, stroking the soft purring head on my lap.

'Look up the McCleery wolf farm. It started in Pennsyl-vania. A pretty interesting place. Anyhow, I loved it there. And when I was small, my mom dropped me at the movies while she did the grocery shopping. I watched a movie all about hunting and outfitting, and I was hooked. It just showed the whole lifestyle, being out in the wilderness and surviving, and I remember them training to be fit enough. It's funny – that small movie changed my life, but I couldn't tell you what it was called. Horses and hunting became my passion, so I moved here to Montana to become an outfitter. It hasn't all been great – I was the outfitter that they sued for $11 million for the 1988 Hellroaring fire, so we had a lot of years of stress, and in the end, I had to do a year's worth of time for it, though it wasn't my fault. I wasn't even there, but I was responsible for the people in that group.'

I watch him now, sitting under an Alaskan sheep on a rocking chair, and I see his face somehow change from young to older as he contemplates the stress of those years. But very quickly he smiles again and the youth infuses his brown eyes.

'Still, we raised good kids. I have an incredible wife – she got me through the times when I didn't know how we could do it, but together we got them through school. It was my family, my friends and my faith that saw me through those dark times. I knew I could start again.

'When I came back, we started horse trainer clinics. You see, we live in America, if you want success and you are prepared to bust your ass, you can have it. I worked with Buck Brannaman, who was the inspiration for the horse whisperer. He is special, he taught me "you don't quit giving". It's why I don't hunt any more. I like the hunt, but I don't like the killing. I became tired of the killing by the end. Now I feel that if I can help people heal the land and help them become sustainable, then I can help heal the planet – and then we won't even have to worry about wolves any more.'

I ask him what he thinks about the wolves here – about living with them, about hunting them.

'It's been an interesting time. I think wolves were always around. Back in the eighties, I saw two of them in Yellowstone, but the reintroduction really inspired so much hatred and fighting. I mean, we all do things we aren't proud of, but I have seen people screaming at each other, hysterical, eyes bulging out. Most people feel this valley has changed because of the wolves. To my mind, the only thing that is constant is change. You have to go with it. The valley has changed anyway. There was a time when I could drive in and see people's lights on, and I'd know who was home and who wasn't. When the wolves arrived, we had to change the way we managed. We had to work on good stockmanship to stop our cows getting killed. You can't change what is already happening, there isn't really a debate to be had – we've got 'em. I'm kind of glad that people could hunt 'em. It helped to diffuse a lot of the anger, it allowed people to feel some sense of autonomy. But wolves are smart – we aren't going to wipe them out, and people need to learn that if you randomly shoot wolves, you end up making more packs. We should all be selective if we want to manage numbers, and shoot a certain amount of pups every year. Some people – well, probably most of 'em – aren't worrying about livestock, though, they are trophy hunting.'

'What is it about trophy hunting?' I ask. 'I get hunting for meat, but I just don't understand what you get out of it when you pull that trigger and see an animal die.'

I'm so relaxed with Vern. He is open and honest, and we are comfortable together, so I find it comes easily to ask him one of the biggest questions I have.

'It's hard to describe, but I guess it's driven by that need we have to collect things. And you need to look at the bigger picture: trophy hunters spend more money supporting wolves than killing them. Sportsmen fund the environment, they buy land, they put in conservation easements, they appreciate the value of habitat, so it helps the state. A guy who is rich will donate disposable income to the protection of the habitat, even if he doesn't live here. There isn't a single hunter that doesn't understand the value of habitat. The wolves just keep growing in number, so there is no need to protect them any more, so there isn't a problem there.'

Vern notices I'm still frowning.

'Look, you have to kill it to put it on the wall,' he says, 'but most trophy hunters would feel remorse.'

I'm surprised to hear that.

'As a hunter, you go through phases,' he continues. 'As a young hunter, you are just amazed when you succeed. It is such a difficult thing to do to track an animal and shoot it right, so you are challenged. As you grow older, you know you have the skill to do it, so you have a choice, and that can be a conflicted feeling for most people. There is all this adrenaline leading up to the kill – will I be successful or not? And then it is over, and there is a feeling of anti-climax somehow. It's hard to explain, it is a primal drive. Like sex,' he laughs.

I laugh too. Vern is undoubtedly making an understandable comparison.

'So it's more about the thrill of the chase?'

'We are hunters at heart. Different people have different reasons to hunt – some for food or hides or teeth.' He pauses. 'But now I can buy a tag to allow me to hunt a wolf, and then I think about it, and then I decide it's not worth the effort. I don't need to take another wolf. It's back to that collection thing – once it is in the collection the hunger for that is gone.'

He looks up at the trophies surrounding him.

'People get mesmerised and inspired by this collection, you know,' he says. 'They want to touch these animals and study them in a way they never could otherwise. See that big tom over there?'

He points at the stairwell. Mounted on the wall is what looks like a rock, and stalking over it is a mountain lion. Its posture is so real, the taxidermy excellent, like a snapshot of lion life right there on the wall.

'That's a story. I shot that twenty-six years ago in Big Crick. I knew there was a big tom working that drainage. I saw his tracks on Christmas Day, but I missed him because my wife said no hunting. Then one night we had the Siberian Express coming up towards us, and I knew it was going to be -40°C that night. I had to get horses in and round up cattle. On my way out, I saw big tom tracks crossing right across the road. They were fresh, but I said to myself "I'm not going to dick around in this, I'm not going to go hunting in this weather," and I drove on by. But by the time I had got everything done, it was just too tempting. I thought "I've got to go catch that lion." So I got the dogs and turned them loose up the track, and we got him up the tree. He was big, you can see, around 170 pounds. There was a wall of white ahead – that snow was coming in fast – but I was warm from euphoria. I didn't get in till eight o'clock the next morning, but I got him.'

'And how does it feel to kill it? I mean, do you shoot them in the head?'

Vern looks at me with a peculiar expression on his face and a half-smile.

'No,' he chuckles, 'you shoot 'em in the heart rather than the head if you want them for a trophy.'

'Oh, of course!'

'At that point, I'm making the conscious decision that is the animal I want. It is more gratifying if you really want something for your wall, for your collection, than letting it live because this truly is a renewable resource – they have protection. I can't shoot it if I don't have a tag to do so. That was a tough winter. I remember some of the elk and deer were on their last legs.

'I can't see why people can't enjoy wildlife for different reasons. If you want to stop people from doing what they

want because of your emotional state… I mean, if we took
away hunting it would leave a hole in the system. There
would be less reason to respect and preserve. Sometimes
being the predator in the system gives people an experience
of the system.'

'But why not just take a photo? Surely then you still get
the thrill of the chase? And nothing has to die.'

'Well, I guess that is where I'm at now. Everybody comes
at the wilderness differently – fishing, hiking, hunting,
ranching, photography – and I don't want to label someone,
'cause people change. Yes, there is the thrill of the chase,
that instinctive cat-and-mouse, and I guess it's about
fieldcraft but also about power. When you are shooting a
trophy, there are so many things that you get close to. It is a
different feeling when you are walking through the woods
with a gun rather than a camera. It's hard to explain, but you
have the potential to finish the deal with a gun. With a
camera, you are looking through a different lens, literally.
It actually takes a lot more work to get a good picture, and
not everyone wants to do that. It's way easier to kill something
than get a good picture of it. Way easier!'

I glance up at the large moose above me. It seems to be
staring me directly in the eye. Below him on the piano,
reflecting the blaze in the hearth, are a set of framed family
portraits. I stand up and stroke the lion rug on the wall. The
head is stuffed, and although the large nose is wrinkled in
a snarl, the eyes are glazed. I notice it has tiny ears. I'm a
wildlife fan, so I can't help but be curious. I do want to
touch it – I can't resist. I marvel at the symmetry of the
markings; a dark line runs from head to tail, and what would
have been the belly is pale and fluffy.

'You can take a picture, but it's not quite the same, you
can't touch it. It's the only way to bridge that gap, to be
able to actually hold the animal. You just can't do that with
a photo. But sometimes the anticipation is better than the
kill. You know, there are only six tags issued for a lion – it is

really such a small percentage of the population here. Montana state manages all that.'

Now that I don't feel embarrassed to study the walls, I realise there are a lot of photos up there too.

'You see that one? Do you know what that is?' I do see it, but it isn't an animal I have ever seen. 'It's a cross fox.'

Before I can ask what he was cross about, there is a knock on the door. Two girls in perhaps their twenties have come to ask if they can discuss the river and watershed issues that the community faces. I realise how connected to the natural world everyone is here. Vern introduces me as 'wolf lady'.

While they chat, I have a chance to examine the grizzly bear looming beside the fireplace. He sits on rocks built into the log storage but reaches his long, magnificent claw out into the room. He doesn't look aggressive, though; his expression is almost benevolent, his glass eyes kind and smiling.

The girls, realising we are busy, promise to stop by later.

'That bear came from Alaska,' says Vern. 'I was hunting up there on the Yukon River with my friend Virgil, and a guy who lived out in the middle of nowhere kept getting his cabin destroyed by the same bear every year. We knew that he was on an island in the middle of the Yukon, which is around half a mile wide there. We had just had fresh snow, and the river was frozen, and we could see his tracks crossing over to it. The snow was ten feet deep, so it wasn't easy to move around, but we got onto the island and started to walk around it. Man, it was so overgrown. It was as thick as snot in there, so quite quickly it got hard to tell where you were. I was going in circles, then I realised that his tracks were on my tracks and my tracks were on his tracks, so I got really nervous. I thought the best thing to do was to sit down and be still for a bit.

'As I was sitting there, I could hear the bear breathing. He was really close. In fact, he was sick of me chasing him and was getting ready to ambush me. I had no idea where Virgil was, and I couldn't hear him either, but I didn't want him to suddenly come across this bear. After a while of sitting in the undergrowth and trying to work out where he was, I realised he was staring right at me. I could see his eyes through the bushes, and I could see his throat. I put a bullet into that throat, but it didn't kill him. I ran off, trying to get back to my trail because now it was just injured and really mad and thinking I had him cornered. He started snapping trees. He was a big bear. I still had no idea where Virgil was, but I knew he would have heard the shot. I had to empty my gun – it took seven shots before he finally lay down and died, and I was so relieved. And guess what? Just as that bear died, Virgil stood up out of the bushes just fifty yards to the left of me – he had been taking a shit the whole time.'

We both laugh, and it brings tears to Vern's eyes.

'How do you simulate that experience with a camera? Would it be the same if you took a picture of that snout and walked away? Also, the poor guy was getting his cabin wrecked by this bear every year. There is this sense of adventure, you see, of camaraderie that you don't get in any other situation, and sometimes it is life and death. The person who hasn't hunted hasn't ever experienced this kind of stuff. There is tradition, too – sometimes families have been hunting together for years, and that is how they bond. I just grew up loving it. I loved hunting birds when I was young – pheasant and duck, and we ate them too.

'Dogs are a big part of it for me. I guess it's everything, the whole lifestyle, training good dogs and good horses, fieldcraft, challenge. It's funny really, a kid from Pennsylvania sees a movie and ends up being a backcountry outfitter – it changed my life.'

Outside the sky is blue and welcoming, and Vern tells me to come and meet his dogs.

Next to a barn are beautiful kennels almost as large as stables. Vern slides the bolts and, almost before the doors open, out shoot two border collies – one older, but one still full of puppy – then a brown and white spotted hound with a beautifully dopey expression.

'I lost my best hound on Sunday. She died real sudden. That was the reason I didn't return your call right away. It will take me a long time to get over that.' Vern bends his head and then shakes it. 'She was just beautiful. I had her trained perfectly, and we had so many adventures together. This one is Dobbie. He's a bit out there. He's good but never will be as good as her. Now I have a choice: do I get another one and spend years and so much time training or is it time to give up on all of that? I just don't know. It's a hard decision to make. I've always had these dogs, you know?'

Vern looks up at me, his brown eyes curious under his hat, as if I might have an answer.

'Maybe I'm getting too old for all this now,' he says. 'Is it time to admit that?'

The dogs bound around, full of energy, eager to be out in the sunshine, looking for a job. I notice that the ginger tomcat who had been stretched out in the sun on the porch swinging seat has disappeared.

'When I made the deck, I left a hole for him so that he could get under it whenever he needed,' Vern smiles. 'That cat has had way more than nine lives. It's way overdue.'

The collies come back to us, eager to find out what we want them to do. Dobbie is on the trail of the cat, who is obviously ten steps ahead of the game.

'This one was nearly taken by a wolf,' Vern says, his hand on the older collie's head, gazing into her eyes as she gazes up at him. Her pink tongue flops out the side of her black and white face as she pants. She looks so happy.

'Seriously?'

'Yep, I had wolves come right down the driveway here, and they had her backed right up to the sagebrush over there. She would have been dead if I hadn't chased 'em off on my horse. That's when I started packing a pistol to scare them off if I needed to. We couldn't hunt them then – that was ten years ago – so they didn't see us as a threat, and they were getting bold. It's what I said about living with them to see the whole system. There were wolf tracks here just last week; they are always around. What did Einstein say? "We can't solve our problems using the same thinking that we created them with." We have to help people learn to shift their paradigms.'

We walk around the ranch. I love it. Dark-eyed cows are contentedly eating fluffy piles of hay, wild elk graze in the paddock, horses snort in greeting.

'What do you think of the hunter in Cooke City?' I ask. 'That seemed to be an easy shot, more like sniping than hunting. Where was the thrill of the chase in that?'

'We live in a place where these animals walk through our yard. Is it wrong? He had a tag, he had his gun ready, it was legal. We should mind our own business. We can't bitch at the hunter if he did it right. Leave him alone. This is a democracy; if we don't like it, we have to change the laws. When those wolves first came, they weren't afraid of anybody. So people taking potshots at them has taught them fear. Living with nature is tough. What would happen if we didn't hunt 'em? It was a mess when we didn't hunt 'em, and one thing that hunting does is give them fear, so the conflict is less. Grizzlies have no fear of us, so they hurt people. Wolves are afraid of us now, so they are elusive, and that is the way it should be. Now I don't need to throw rocks at them when they come up my driveway, because they don't bother. This guy in Cooke City was being opportunistic, yes, but there are a lot of people who are jealous of him for having that trophy. He just happened to shoot a wolf that people had put a name on. A member of a pack on the down low and in the wrong place.'

'OK, but think about what that does to the pack, Vern,' I respond. 'That family's future is now in jeopardy because of what he has done.'

'Well, they weren't doing well anyway, and the fact that an alpha female has died is not an abnormal event in the life of a wolf. We just can't make the whole of Montana Yellowstone

National Park. If we couldn't hunt with the numbers rising as they are, it would be pretty hard to live around here.'

A breeze picks up, and in the distance, clouds are brewing.

'We're meant to be branding cattle on our neighbours' ranch tomorrow,' he says, 'but we won't be if it is raining. It will be the 154th branding on their ranch. His great-great-grandad made his money in the gold rush, and the family have been there ever since. They brought a herd of Texas longhorns all the way up from Texas. Those guys understood how to manage cattle. Sometimes they would be under attack by the Native Americans, or wolves would be after the cattle, but they still got them here. A lot of those skills have been lost now, but they still matter, especially when it comes to managing diverse habitats and keeping them safe from predators. Stockmanship,' he adds with a smile.

I drive the potted miles back up the track to the Anderson ranch where I am staying. Vern was certainly not what I expected a trophy hunter to be. He is nice, kind, loves his family, and in many ways, I am aligned with him. He is a good human being. I'm happy that he felt comfortable to speak his truth even though we were in disagreement about some things. He has reasons for the things he does. In his eyes, there is no suffering, and he doesn't take more than he needs or more than is available. Nor is he the type of hunter who would go to Africa to shoot animals that have been bred in captivity for trophy hunters to kill. He is a man of his environ-ment, with a passion for it and a deep respect for wildlife.

Vern's final words are ringing in my head. Not the ones about me coming down to meet his wife for a drink, for now, we are friends, but the ones where he said, 'How can I have time to trophy hunt when I have to save the planet?'

I guess Vern is an evolved trophy hunter, although he might say God designed him that way.

CHAPTER SEVENTEEN

Another way

The Anderson ranch lies at the top of the hill looking over the valley. I have been welcomed here by a family with an interesting take on what it means to live harmoniously in this environment, and again I make a new friend: Malou.

We bond first because we are both working mothers. Malou and her husband, Trey, have two girls living on the ranch with them. Hazel is two and Esme is five, and they are delighted to see me. The first thing they want to show me is Napoleon, who is actually right in front of us as he has been following them around the ranch. Napoleon is a small black and white calf who is only around two weeks old – I'm guessing he has been named for his eye patch.

'Come,' says Esme, wisps of her hair turning gold in the sunshine around her face as she takes my hand. 'Come and see where he sleeps.'

I assume she is about to lead me to the barn on the other side of the yard; instead, she walks me over to the climbing frames, swing and little playhouse, and I wonder if I'm being duped into playing. Little Hazel, shyer than her sister, follows a little behind.

'Come on, baby,' says Esme, holding out her hand before looking up at me and explaining in a more grown-up tone, 'We are playing babies. Hazel is the baby.'

'Ah! I see!'

As we reach a white playhouse with a baby-blue roof, Hazel runs to catch up.

'He sleeps inside!' she says, wanting to be the first to share the news.

Hazel opens the small door, and we all look inside. Instead of the typical row of plastic teacups and a table and chair that you might expect to find, in this playhouse mama has spread a lovely deep bed of shavings and hay.

'She had to put it all in there so he wouldn't get cold. He chose it for his bed!'

The long-legged Napoleon stands just beside us, his eyes large and a solemn look on his face. Both girls throw their arms around him and kiss him a lot.

Next, I am introduced to the pack of dogs. I guess you need a pack if you live here.

'This is Duchy, the puppy.' We all cuddle him next, another long-legged black and white creature but a collie with a freckled face and an eye for mischief. Life for Duchy is clearly one big game, and he wriggles out of our grasp and canters off to eat cow poo.

'That's Simon,' she says and points at an older border collie who is keeping his distance.

'Sage-y,' she calls, and Sage trots over, her tail high. She looks like a perfect mini wolf, and I find out later that she is a Saarloos wolfdog, a line that came from crossing wolves with German shepherds in 1930s Holland. Apparently, they make excellent herding dogs.

'Grandma rescued Sage from Alaska.'

'Wow, she is lovely!' And she is – very affectionate.

'Esra?' Esme calls. 'She's over there.'

I spot a huge white shaggy dog over the other side of the paddock, a flock-protecting breed, known to be aggressive to predators but soft with children and the animals they protect. There probably isn't a more perfect home than this if you are a Great Pyrenees. I think I've actually seen more of them on the ranches of Wyoming than I saw while I was presenting Crufts for the BBC. Esra wanders over, sniffs my hand and decides I'm not going to be of any bother.

Malou is doing the groceries and Trey is in their cabin, so I excuse myself, leaving the kids trying to get the unwitting

Napoleon back in his 'bed', and head over to the old stone house where I am staying.

I wonder what it takes to be a working mum when your work is ranching and not just sitting at a desk all day. Later, as the wind picks up outside, bringing flurries of spring snow, and the evening steals the big sky, the kids are in bed and Malou is ready for that first glass of wine of the evening, she heads over to the stone house. We light a fire in the huge wood burner, settle beside it with a bottle of red and talk.

Malou is the first woman to manage the ranch that has been in her family since the 1950s. Her grandfather bought it. He was a pilot in the Second World War who was shot down over Italy and was a prisoner of war for two years. He was a well-educated, ecology-minded man with PTSD, and this was his sanctuary from all that suffering. He had to learn how to ranch from the neighbours. Now they are the family who have been ranching here longest.

Malou grew up here, one girl among four boys. The ranch sold custom grass-fed beef and they had sheep too. This singularly beautiful spot, which feels as though it is on top of the world, surrounded by mountains and wildlife, was home until she went to boarding school and then headed east to do a psychology degree. Many moons later, she found herself burned out and knew that city life was no longer the path for her.

'But could you see yourself ranching?'

'I needed a paradigm shift to make that happen,' she says as she sweeps her long dark hair behind her ear, tucks her feet up under her and rechecks the large logs are taking the new fire up from the kindling. 'It wasn't till I took a course in ranching for profit in a holistic way that I could see I could do it. That is what allowed me to become a rancher, and I guess that is because our family never really approached things in the traditional way because we never really knew the traditional way. So we were always open-minded and willing to learn, willing to question whether we were doing things right because we always had done that. Anyhow, I was very comfortable with this more regenerative model I was learning about, and so were the rest of the family.

'Dad had really done that all along. And so when the wolves came, instead of fighting it, his philosophy was to embrace it – this change is going to happen. We actually got rid of all our sheep at this point. He was the only rancher that felt that way. But – and this is important – he had another form of income. He and Mum were both working, too, and so that made a huge difference to the way we viewed things. We weren't entirely dependent on the ranch for our living, we had to supplement it. So we moved to a business where we sold cows and calves, and we ran yearlings.

'For me, this is about our community. And when I say that, I don't just mean the people, I mean all of it: the wildlife, the soil and the water, the mountains. We have a unique set of challenges neighbouring Yellowstone as we do, and wolves are just one of them. We have bears, wildfires, extreme weather, tourists – so many challenges. One day my sister-in-law Hilary, who is a biologist, and I were talking about all of that, and we decided to set up an association. Our mission, when we started, was preserving the rural west, trying to understand and work with this wild place. We had similar viewpoints but came to it from different angles – her with the biology and me with the ranching experience – so we made a good team. We were really inspired. We wanted to keep the good traditions alive, and we wanted to maintain an intimate connect to the landscape to provide good stewardship and be taking care of the soils and water. So much of our history with our habitat has been about ego, a kind of heavy-handed male domination that destroyed too much.

'We first got some funding from Defenders of Wildlife, so that set us up, but we now have funding from both liberal and conservative sources, which I'm pleased about because otherwise you just get labelled as a bunch of hippy wildlife people – which I hate. Now our mission has developed a little, we are about supporting diverse wild and working landscapes, and fostering good and trusting relationships between ranching families who live here. Hilary started the range riding programme up here, which was really interesting.'

Malou yawns and I catch it, recognising the yawn of a mum of young ones. The wood burner has done its job; the room is now soporific.

'Perhaps we can talk about that next time. We have no rush, I'm here for a while.'

'Absolutely,' Malou laughs. 'We were all ill last week, and it's taking a while to get my energy back, and the girls are pretty full-on, as you know.'

'Let me help! What do you need to do tomorrow?'

'Well, we will be out getting groceries and then we have to rescue Annaliese.'

'Who is Annaliese?'

'She is our beautiful Jersey cow who is meant to live in this paddock next to the cabin right here, but a couple of weeks ago she aborted her baby, which was really sad. Then, a few days after that, she took off, and we found her two miles away on my brother's farm. She had literally leapt over fences and was in with the bulls! So now I don't know if she is pregnant or not. But my brother managed to get her out of there and into the paddock with the mums and babies. When I went to see her, she was cuddling up with any of the mums who had newborns. Well, that was what I thought at first. But then we realised, of course that what she was trying to do was steal a baby for her own,'

'Oh! I guess her instincts took over,' I say.

'Yes. I can't imagine what she was feeling, that drive for a baby was so strong. They make great mothers. What she doesn't know is that we now have a two-week-old calf here for her. I just have to persuade her to come home and then she will get to meet Napoleon, and that is a job for tomorrow.'

'I'll help you.'

'Great.'

We both smile as we finish our wine, knowing it probably won't be that easy.

CHAPTER EIGHTEEN

Wolves up the arse

'They stuck wolves up our arse.'

I chuckle. This hunter, Ron, didn't have the time to meet me but was more than happy to talk on the phone, and he doesn't hesitate to get right to it. I barely get a word in.

'There were wolves there before the reintroduction,' he says. 'We didn't need any more. Don't get me wrong, I'm not a wolf hater. They are a cool animal, a big predator and, they have the whole "hierarchy of the pack" thing. They're as cool as shit. No, I'm more anti-government than anti-wolf. The wolves took down the elk population, which we historically hunted. I have hunted wolves in Alaska, I have some trapper friends that caught two here, and I buy a wolf tag every year, hoping to see one and shoot one. We need a better balance. We need more elk, but those wolves are hanging out on private land so we can't shoot 'em. I've lived through a bygone era that we will never see again, with 20–30,000 elk herds – that's over now.

'Yes, I think wolves are as cool as shit, but yes I'd like to get one or two a year. I respect animals, I don't want to decimate them, but if there is a harvestable number I would like to be part of that. I like to hunt and trap 'em. We match wits with one another, and we outsmart them. That is the challenge. These wolves here are cagey, they are wily and smart, and I only saw one last year because of that.'

'Would you trap a wolf?'

'Yes, I would.'

'But what about the suffering?'

'The suffering doesn't bother me in the least. I've seen how cruel the animal world is and just how much suffering there is. Wolves kill just for sport, not just because they are hungry. In Alaska, I came across twenty-one Dall rams killed, and they only ate parts of two or three and never came back.

'Mother Nature is really cruel, we have a hard time understanding that, especially people who haven't lived on the land. I know what coyotes and wolves do to other animals. The pain and suffering of trapping have nothing to do with it. I'll do trapping demonstrations for kids at schools, where I will put my arm in a trap and walk around with it so that they can see it isn't that tight. You slowly lose feeling, it gets numb, the trap cuts off the circulation, it isn't near as painful as what the antis would lead you to believe. Wolves eat things alive. When we find something in a trap we put a bullet in them, and they die – it's quicker, it's not as cruel.'

'What kind of trap would you use?'

'There are several traps – a foot trap for coyotes and wolves, but there is also a cable snare that will catch them around the neck.'

'That sounds horrible.' I can't help it.

'I feel good about it. I feel that it is a balance. Trapping is a management tool. For every coyote I trap, I save hundreds of other animals that predator kills. Take my son who has a remote cabin in Alaska. They had a government who was anti. My son has that cabin for premier moose hunting, but when the wolf hunting stopped there were no moose left. That's why Sarah Palin instigated a predator-control programme. There has to be a management balance. There are too many of us in lots of parts of the world. If you have too many of this, you have a lot less of that. Their numbers need to be controlled.'

'But it seems to mean more than that,' I say. 'When I see people say things like "the only good wolf is a dead wolf", it seems there is a hatred for wolves that isn't there with other predators.'

'That's because they are and always were competing with humans. We are against the wolf because they are competition, especially with the early settlers, the cattle ranchers and the hunters.' His tone softens. 'I get that it is hard to understand. My wife is the same. She had never hunted, and she shot her first elk, and afterwards, she was crying and petting it. When I kill an animal, there is a part of me that is sad for taking a life that the Lord created. It's not a bragging thing for me. The good Lord provides opportunity, and we should take it seriously. It comes down to a lot of things – how you are wired, how you are brought up. On a ranch, you deal with life and death all the time. That was my childhood. It doesn't mean you get hard and callous.'

'So what about the wolf that was shot in Silver Gate? What about taking out a member of a family, one that is needed and mourned over?'

'The next season, they will be right back to being a family. It depends on the pack. Next year the pack goes on, goes back to the same denning spot and starts again. The impact can be small. Wolves are pretty capable of surviving and moving on and dealing with a situation that can happen naturally. It just isn't that big of an impact, I don't think.'

'Yet it causes a big outcry. A lot of people care and don't want this.'

'Some of the first people who harvested wolves got a lot of bad attention, which I don't understand. I mean, I don't mess with their lifestyle, so why do they think they can mess with mine? Any animals need to be managed. We aren't going to decimate them again. I mean, in those days when they were out to exterminate every last wolf, they were using poison by the end. We won't ever go back to that. Wolves are doing well now, so why do these people have the right to tell me that I can't harvest one in a managed situation? They put their money into campaigning and complaining and their energy into posting all those negative comments, but when you look at it, hunters are the true conservationists who really put their money where their mouth is.'

By which I assume he means keeping the habitat good for wildlife, and the investment that takes. Ron is winding down.

'We have to take a surplus off. I don't want to get rid of all wolves, but hunting is a tool, trapping is a tool. I believe the good Lord put the wildlife on this earth for us to enjoy. I love it. I don't want to see it go in my lifetime. I try to go hunt or trap or backpack every day. That is never going to stop while I'm on this earth. I love the wilderness; it's why I'm here. I hike to the lakes, I go fishing, I see mountain goats, wolverine, lynx. I live to be in the outdoors. I do hunt coyotes out of a plane – you can't do that with wolves here in Montana now, but I would.

'Anyhow, I hope I've helped you with your book. It's been a pleasure talking with you. Sorry we didn't get to meet up.'

And with that, Ron was gone. In one brief phone call, I've heard his position and his reasons.

I think of Lisa Robertson hunting wolves in a plane for a very different reason. I wonder how us humans will ever be able to meet in the middle on this.

CHAPTER NINETEEN

Willow Creek ranch

The idea that hunters invest in conservation is fascinating to me. They are adamant that this is the case. I'm wondering about it as I drive through Livingston, Montana because I'm hoping that today I will find out more about it.

Livingston is a busy, pretty town, full of Americana, on the Yellowstone River and a little more than an hour from the park. It has a cool history. It started off as a trading post because of the river, and at that point, it was called Benson's Landing. Then, thanks to the arrival of the Northern Pacific Railway, it proliferated. It was called Clark's City briefly, after one of the contractors, but then they decided to move the whole place three miles to the north, closer to the railroad and rename it Livingston. It became a stopping-off point so that engines could be serviced and attended to before they ascended the Bozeman Pass, which was the highest point of the railroad. I'm not surprised to see that the rail tracks, engines and massive wagons are still weaving their way through the town. When a spur line was built to run south down to Yellowstone National Park, Livingston became known as the original 'gateway' to Yellowstone, bringing valuable tourists from the east.

I can feel the history here. It wouldn't take much to turn back the clock, to get rid of the new supermarkets and gas stations and see the settlers and railroad builders creating many of the buildings that still stand.

After leaving town, it just takes a few minutes to reach ranching land and what begins to feels like wilderness. As I follow the hastily written directions to meet my next hunter, I am really surprised by how close the wilderness comes to the town, and delighted by sweeping Wild West

views, the gentle angles of the auburn and tan plains giving
way to a distant mountain range. It is easy to understand
why settlers thought this was some kind of promised land
as they claimed their lots here.

Ed is waiting for me, and I see him waving as I drive
under the wooden archway that welcomes you to the ranch
and marks its boundary. He's managed this ranch for ten
years and knows it well, and he's offered to show me around
so that I can see a different kind of ranching. This ranch is
not managed for profit; the owner is wealthy – he doesn't
need to run the ranch as a business – it is managed for
wildlife and as a legacy for his family.

Ed greets me with a big smile. He is smaller than me,
with a broad, red face. He looks like a powerhouse, and
when he takes his sunglasses off to say 'Hi! Welcome!' his
eyebrows are so blond I can barely make them out under
his camo cap.

The minute I walk into his home, I realise he is also a
trophy hunter.

A huge Dall sheep trophy takes pride of place in the
corner of the living room. By now, I have spoken with
enough hunters to know that they are a challenge to hunt.
They live in mountainous regions in Alaska and western
Canada, and they are rugged and intrepid – the kind you
find grazing calmly on a cliff edge. This one is indeed a
beautiful specimen, off-white with beautiful amber-grey
curled horns that take years to grow into circles like that.

'That trophy, those horns, they are what clicks the
memories,' Ed says. 'It's something to share with people. It's
the hunt I love. The kill is the point, but you don't go
through all of that just for that moment. Sheep hunting is
especially gruelling – that one was a thirteen-and-a-half-
hour stalk in really tough conditions. I think "trophy" is a
bad name for it, but it shows your accomplishments, and it
reminds you in a way that nothing else can.'

'Kind of like the ultimate souvenir?' I suggest.

'Exactly! I've hunted 'em all – grizzlies, caribou, mountain goats, black bear and deer. I've been bird hunting. Hunting gets you out there. The things you see are amazing. I do a lot of videos too and, as I get older, probably more photos. It's a passion. It's something you've grown up doing so you love to do it. You get addicted to the adrenaline. Everything I have ever harvested I feel for, and you have to have respect for it. Not everyone does it, I know, but I will say a prayer over each body.'

I find myself staring at the sheep. It's beautiful undeniably, but I still can't imagine what it must be like to be the one who pulls the trigger and watches the life fade.

'So,' he says, 'do you want to see the ranch while the sun is still out?'

'I'd love to!'

The Gator is waiting for us outside and, although the sky is blue, the wind is biting, so I am relieved that it has a warm cabin.

'I prefer this to my car,' says Ed as he starts the four-wheeler's noisy engine. And I can see why – it feels so sturdy and small, as though you could literally go anywhere in it.

Ed tells me that his two-year degree was in forestry, so I immediately feel that we must have a considerable amount of shared understanding. After all, to be a forester is to understand ecology. After five years in corporate America, he went back to school at the University of Montana to study resource management and continued with his passion for forestry, which he did as a job after that for many years before deciding that he wanted to try something different. So in 2009 Ed took a job managing Willow Creek ranch and every single one of its 19,000 acres.

'At that time, it was a straight cattle ranch,' he says, driving us around the barn and towards a track heading off into the distance. 'Originally the land was about ten smallholdings but Harms cattle company bought them all, and they raised cattle on it and then they had to sell it to pay inheritance tax in 2008. Now I manage it for wildlife and for recreation. We do have some cattle, and we lease out some grazing, but I use the cattle as a tool. We aren't interested in making money out of cattle here – we use them for grass-management purposes only. We only ever graze half the ranch maximum, and we have a two-year rotation. By managing it that way, we have a much greater diversity of plants, and we have tripled the wildlife because we have increased the habitat and provided plant diversity. So we have alfalfa growing here and high-protein forage, some wheat, some oats.'

We stop for a moment. A large herd of mule deer are hanging out on the gentle, grassy slopes to our right. Some are lying down and enjoying the sunshine, others graze, and a couple of them lift their heads to look at us. Mule deer are greyish-brown with white bottoms and pale bellies. They are pretty average deer really but for their massive, dark brown ears, which are way out of proportion and dwarf their faces. These ears can twirl around like aerials, and frankly, I find them hilarious.

Beyond the deer, the slope steepens, and the land becomes rocky and interspersed with juniper trees. It then changes to cliffs darted with vertical lines, and they surround us, sheltering the valley. It is a beautiful spot.

'We have more mule deer, antelope, elk, bears and lions than have been here in many years,' Ed says.

'Does that mean you get wolves here? It looks like wolf paradise to me.'

'We don't get many – and to be honest, I'm happy about that – but as a matter of fact, we found a wolf track this morning right outside the main ranch house.'

'Can we see it?'

'Sure. I'll show you, but we'll go the long way round so that you can get an idea of what we are doing here.'

He starts the engine again, and as we head up the track, a pair of white-tailed deer startle and take off up into the hills. They too are hilarious. Their long white tail lifts as a warning and wags high in the air as they speed away. I find it odd that a prey animal would evolve to make it easier for a predator to keep sight of it during the chase.

'These were old calving pastures, and now we have converted them to their natural state,' shouts Ed above the noise of the engine.

More mule deer find our presence startling and leap away, all four legs leaving the ground as they jump high and fast.

'Do you know what that's called?' asks Ed. 'When they leap high off all four feet like that it's called stotting.'

We are bumping along beside a narrow creek to our left that winds, crystal clear, through the willows.

'We've cleaned up the creek, allowed those willows to grow so that the beavers can come and the banks are full of wildlife. We get moose and black bears in these willow stands all the time. Although the moose numbers have stayed low since the wolves came back.'

'So do you get trout in there?'

'Yup, and we have these pools, so we have cutthroat trout in there.'

I notice a large pool above the stream. Then Ed pulls over again.

'Take a look at this, I want you to see this. We have replaced most of our fencing now with wildlife-friendly fencing.' We clamber out and inspect the fence. 'See, ordinarily, in traditional fencing, you will get a T-post every sixteen feet. That is the post that keeps the whole thing standing. Then you'll have four strands of barbed wire – the first strand is around one foot high, and the top is between forty and fifty inches tall.'

'Right. So what is the difference?'

'Well, here on this fence you see we have a T-post every fifty feet. Then we only have three strands, and the first strand is up taller at eighteen to twenty inches, and that means the calves can get under it. You see, when a herd of elk comes across the fence and the mamas jump it, the calves have nowhere to go. This way, they can follow mom right through.' He stops talking and leans right into the fence, which bends with him. 'And that's where having the T-posts so far apart really helps. The whole thing flexes so the elk or deer can't get twisted up in it and trapped. It's a win-win situation: the elk don't get caught or injured, and for us, it's great because they're not tearing up the fences and so we have way less maintenance, and maintaining fences on a ranch is normally a major part of the work. And it is cheaper to build – not by much, but it is still cheaper.'

We get out of the wind and back into the car.

'Six years ago,' Ed continues, 'we started putting them in, and they have been great. It isn't that elk are stupid, though. When they are calm and just moving around, I have seen them go miles out of their way when they have calves, just to go through a gate they remembered rather than risk a fence. In those six years, there is only one place where we had to fix it and only once because an elk cow got stuck – they were in such a panic because they were sprung by wolves. We knew that because an elk was dead in the middle of the field.'

'How did you know it was wolves?'

'Well, it wasn't where a lion would kill, and the elk was hamstrung. They ate some but not all of it. Perhaps we disturbed them.'

'Right.'

The Gator climbs the steep hill ahead of us with ease. There are still patches of snow here and there as we get higher and higher until we are right on top, looking down at the flats below.

'See that over there? That's a buffalo jump.'

'A what?'

Ed is pointing way out into the distance. The sky is big and blue with a few whipped clouds, the flats are just turning green, and on the horizon is a line of hills that ends abruptly with a steep cliff.

'It's where the Native Americans would herd the buffalo. They would get them to stampede right over the edge of the cliff, then they would have their legs broken, and the other tribe members would be waiting at the bottom to kill them. People have found a ton of bison bones down at the bottom there.'

Looking at the shape of the hill, it is easy to imagine the buffalo leaping off the top. It must have been an incredible, if weird, sight to see. Apparently, it was a hunting method that started over 12,000 years ago. And they say there isn't much history here in America.

'From here you get a good lie of the land. The elk spend a lot of time on the flats, grazing. So if the wolves come, they would come on this flat to hunt the elk but live up there on the hill in the timber. You can see we have three drainages.' Ed points out three ravines interspersing the hills. 'That over there is the north fork and then the south fork and then Ferry Creek. We think when wolves come down, they use that north fork because that links right up to the Beartooth mountains where they live. And I think there is another pack just north of here in Bangtail Creek.

'We don't really have a problem with wolves here. A pair came here every so often, and they both had collars on them. We haven't seen them for a while, and they have never bred. I don't know why. Then we'll get the occasional single male running around like today, travelling through the open country. Like I said, we have only had one bull elk killed. The mountain lions kill more here, and we lose mule deer yearlings to them and the coyotes. Elk and deer aren't used to wolves here, so if they do show up we know it because they disappear – they behave differently and stay away for a few days. We have noticed the coyotes leave, too, when the wolf tracks are down, and that isn't always a bad thing because they are really tough on deer fawns.'

The cold wind whips my hair, but the view of the Wild West is too beautiful to walk away from.

'Lewis and Clark camped out somewhere in this valley. We are pretty sure in the book they are talking about Ferry Creek, but these aren't Lewis and Clark days.'

'What do you mean?'

'We can't just let the animals be, like the far-left side think they should. You can't go back to the way it was then, especially if your house is where a herd of buffalo used to be. Also, wolves enjoy the hunt as much as the feeding, and they would rather pursue the elk. In the Seeley Lake area where I used to work – the Bob Marshall Wilderness there – we had so many elk. And in the Blackfoot Clearwater game range, look at the numbers: there were something like 14,000 elk and now just three hundred, and there are lots of wolves there.

'We have to manage the situation. We are manipulating nature – have we manipulated it too much? Probably, but in Montana, there are way too many wolves. When they reintroduced them, their goal was not nearly this many. Yup, we have too many wolves and too many packs. I think there are probably around five hundred wolves around here in total now. How many wolves were killed by Fish and Game because of livestock predation? Yet the numbers are still steadily growing. The hunting population hate it. The moose population was hit first, but then the elk population went right down. In 1994 there were 24,000 in Yellowstone, and they are now down to 7,000. Hunting hasn't changed, so it's the wolves taking them.

'It's so complicated, the ranchers have to deal with elk eating all their hay and grass, and antlers on the ground giving them a flat tyre on the tractor. You would think they'd like a wolf to take down the elk. It's the hunters that really hate them, they want them out. Some people depend on hunting elk for their livelihood – if they run an outfitting business, you know. Those are the ones that have really suffered. Is it cool that people can see wolves in the wild now? Yes, but it's getting out of hand here in Montana.'

'What would you do if you saw that wolf here now? Would you kill it?'

'Hunting season is over.'

'OK, what if we were still in the hunting season?'

'Personally, I'd love to get one. I've never shot a wolf before, and that first one would be more for a trophy and from then on for management. I have two sides of me: the rancher that needs these animals to be managed and the balance to be kept, and the hunter. I like to add to my collection, but once I have a trophy, I have it. If I have a black bear, I don't need another.'

There is no sign of a wolf up here, just the herd of elk massed together on the next hill. We head down the hill and into the valley ahead.

The wolf tracks are beside a stream next to the ranch house. The ground is muddy, the tracks fresh and really big.

'I wonder where he is,' I say. 'He could be watching us right now from those willows.'

'Oh, he'll be long gone by now,' Ed replies.

Hunting season or not, I hope so.

We round off the tour by driving through a herd of black cattle with small babies.

'So, none of the cattle have ever been attacked?'

'No, we have been fortunate.'

One thing has been bugging me, and I can't quite put my finger on it.

'Do you hunt on the ranch?'

'The owner does, and we take out a few outfitting parties through the year.'

The penny drops.

'So Ed, tell me, when you say that you manage the ranch for wildlife, what do you mean?'

'I mean for elk and deer and such.'

'Not for predators though, not for the whole ecosystem?'

'No, this isn't Yellowstone. We have the predators here, but we have to keep the balance right for what we need. If we had a problem with wolves or too many bears, we would need to deal with it.'

'So aren't you managing the ranch for hunting rather than wildlife?'

'In a way. Like I said, we manage it for wildlife and recreation.'

'I see.'

And I think I do. Here, hunting is being part of the wilderness, it's what a lot of people do and have done for generations. For the hunters I have spoken to, killing animals – whether for meat, for the challenge or to manage numbers – is something that is as natural as the natural world, and while they are permitted to do it, they will continue to do so. This is not something that will change quickly.

On the way back up through Tom Miner Basin, I pull over to take in the view. The Yellowstone River meanders across the plain in front of me, snow-capped mountains are ahead, and I am high up in the hills. I am deeply in love with the landscape of the Wild West; it never ceases to move me. I take a moment, making my mind blank and just drinking it in.

A truck passes and pulls over in front of me. A large man gets out and walks up to my car. I open my window.

'Hi!'

'Are you OK?' he asks, another thing I love about the Wild West.

'Yes, I'm just taking in the view.'

'Gorgeous, ain't it?'

'It is.'

We contemplate it in silence for a moment, and eventually he sighs.

'Well, I just wanted to check that you didn't need any help.'

'No, thank you, I'm good.'

'Doesn't sound like you are from around here?'

'No, I'm from England originally, but I live in Jackson Hole.'

'What brings you here?'

'Wolves,' I smile, wondering what the reaction will be this time. 'I'm writing a book.'

He looks interested, 'Oh!'

'You?'

'Same,' he says, 'only I'm not writing a book. This is my old stomping ground. I moved to Bozeman for work, but I still come here whenever I can to hike around and see some wildlife. I used to work on the ranch up there.'

'Have you seen wolves today?'

'No, I'm finding it hard to see them this year, but last year I did. They are getting better and better at keeping themselves hidden. They are pretty good at that when they want to be. I hiked up that hill over there, looking for them when there was still a lot of snow on the ground. I spent hours looking and when I came back down, there were wolf tracks all over my tracks, but I never saw one wolf that entire day.' He laughs.

'Did you see them when you worked on the ranch?'

'No, but that was because they had only just been reintroduced into the park. I was just casual labour, you know. I spent most of my time out fixing fences, but they told me the minute those wolves come over that ridge you shoot, shovel and shut up. That was before it was legal to shoot one, but they were really worried up here, and that rancher said to me "you shoot it, you bury it and you say nothing to no one.'"

'And did you?'

'No, I was lucky I never saw one. I was glad. I don't think I could shoot a wolf.'

'Me neither.'

After another sigh, he says goodbye and leaves me with some parting words: 'Watch out, the bears are up.'

It seems like everyone who lives on the edges of Yellowstone and Grand Teton Parks has a wolf story to tell, and usually a bear one too.

CHAPTER TWENTY

Will the cow continue to live in the playhouse?

Annaliese does not want to come home and is very tentative about approaching the bucket of grain Malou is holding for her. She suspects we are trying to bribe her. Malou and I look at each other, shrug and smile. She may be a cow, but she ain't stupid – we are definitely trying to bribe her out of the maternal cow paddock so that we can get her home. She has decided, however, that she most definitely doesn't want to leave this paddock. The very next calf that drops could be hers.

We watch her try to mingle back in with the others. This field full of precious mums and babies has its fence surrounded by another, electrified one, but not the type that we usually see. This one is hung right around by orange flags: fladry. I'm not surprised they are using it here as it was designed to keep wolves out. The flags are off-putting to them – not worth the challenge.

'How is the fladry working for you? Lots of ranchers seem to be sceptical about it.'

'We have had a 100 percent success rate with it for eight years. This is the most sensitive herd in the area, the only cow-calf herd in the basin, and there are no elk or deer babies on the ground right now, so they are prime prey. You have to use it right, though – not too much that they get used to it, and you have to enclose the circle and have it electrified. We take it down almost immediately after the last cow has had its calf so that the wolves don't get used to it at all, and they are still frightened of it. When the calves come out of fladry with the cows, it is a really good time to re-teach them all to bunch up and be a herd to protect themselves. So that's when we start range riding with them, which will be about mid-May. We have visual data on bears and wolves checking out the fladry and running away. Electricity is also extremely important for bears in particular – fladry is far more effective combined with electricity.'

We contemplate Annaliese while she pretends that we don't exist.

'This is part of my family's ranch. That little building there,' Malou says, pointing over to a simple one-storey log cabin with a traditional porch, 'that was where Grandma used to live. One night she left the bedroom window open, which we would never do now, and she had a visitor. At the age of ninety-two, she chased a grizzly bear out of her home, but he did a lot of damage in the meantime. Grizzlies are just as curious as wolves – if not more so.'

It may seem like a simple life up here, I think, but in truth, it is anything but.

'Annaliese doesn't want to come home,' shouts Esme from where she is sitting on the fence beside her little sister. They are, without a doubt, the cutest couple in the valley right now.

'I know, honey. You guys just stay there out of the way.'

'She doesn't know we have Napolean at home waiting for her,' Esme calls.

'Annaliese,' shouts Hazel. 'Annaliese! We have a new baby for you… She's not listening, Mama!'

'I know, honey. Can you guys get me some more grain?'

Malou starts dragging a large gate. I can tell she is still feverish and struggling with the flu. Just the physicality of ranching is tough. I join her, picking up the other end. It's heavy, and we sweat to set up a system of closed gates, forming a nice one-way funnel leading to a small corral. We just need her to be greedy enough to follow us in. From there leads a chute, and surely it will then be easy to persuade her to get into her trailer, which is parked at the end.

At least, that is our plan. We have only managed to get her as far as the muddy swamp area by the gate when she kicks up her heels in a coquettish fashion and skips back to the breastfeeding herd.

'As soon as we get back to the ranch, that cow is being halter-trained,' says Malou through gritted teeth.

'I'll help the girls with the grain.'

It takes a lot of time, patience and grain to get Annaliese to do our bidding. At the last minute, I have to sprint when she spots a cow-sized hole under the fence just as we are about to close the deal. I fend her off. Then there are a good thirty minutes of persuading her to clamber into the back of the trailer and keep the girls safe at the same time.

'I'm so glad you are here,' Malou says. I'm not really doing anything helpful, but it is always nice to have a second pair of eyes when you have two small kids – that bit I understand.

'I'm trying to persuade her rather than force her. It's better to manage cattle this way – it keeps them calmer. Girls, don't climb into the chute, please. Stay up on the fence. It's called low-stress livestock handling, and it implements everything we do. When we herd them, for example, we use a zigzag technique so they can see us out of the corner of their eyes, and that inspires them to move in the direction we want them to go. The alternative is to go behind them, shouting and yelling, which gets them anxious and kicks in their prey instincts. Calm cows are safer cows, not behaving out of anxiety or fear, and that also leads to better beef production. What we are trying to do is look at things from a different angle – instead of "how can we get rid of predators?" it's "how can we reduce cattle's vulnerability?"'

Annaliese turns around again – no, she is not going into the trailer.

'She isn't normally this difficult. She's normally really placid. It's one of the reasons I wanted to try a Jersey – for their temperament.' Holding her hands on her hips, Malou stops for a breather.

'I'm sweaty. Can you take off my jacket?' A little voice down to my left and a small tug at my shirt.

'Of course.'

Freed from her jacket, Hazel climbs back up behind the chute. These small girls are watching every move their mum makes, learning in a classroom surrounded by mountains and in a valley with a resident pack of wolves and the densest bear population in the lower 48. The sun is high and hot. Because we are at altitude, the weather is, as ever, extreme – a bit like Annaliese's mood swings.

'I think she is still in heat. I don't think her frolic in the bull field went to plan.'

'How come?'

'All this moodiness, and when I walk behind her, she shifts her tail to one side as if she might be expecting something.' We laugh. 'I'm quite glad. I didn't want her mated with one of those bulls, and hopefully when she gets home and meets Napolean that might calm her down a bit.'

Malou clambers into the back of the trailer, fluffs up a fresh pile of hay so that it looks its most enticing, and rattles the grain bucket again. Annaliese looks on. A few moments later she decides this game is boring after all and she really has no reason to stay now that she is separated from the calves. So – calm as you like – she steps daintily into the trailer as if she had fully intended to do that all along. We keep quiet till the trailer gate is shut and then we all cheer.

'You did it, Mama!"

Annaliese calmly chews on her hay, waiting for her ride home. This is the life of a working, ranching mother.

Back at the ranch, after an excited Annaliese settles down with Napolean, we collapse on the deck at the back of the cabin with a cup of tea. The girls pootle about, enjoying the new-found freedom of a sunny spring day, rediscovering summer toys. They decide to set up a toy farm on the deck next to us.

'Tell me more about the Tom Miner association. How does it work?'

'There are really five upper basin ranches here, and one of the things we have done is set up a joint range rider programme that three currently participate in, but that does fluctuate. But we do other stuff too – carcass research, data collection. We collect data about the wolves, bears, livestock and deer, but we also do things like monitoring the effects of drones on those animals so the FAA can make guidelines for drone use.'

'Really, is that a problem?'

'More and more. Not one we would have predicted, but of course, drones are not allowed in the park, so people come to use them in the neighbouring areas.'

Hunters of a different kind.

'So what does a range rider actually do? I mean, it looks like the best job in the world to be out there on horseback, but are they actually there to scare off wolves? That might get a bit iffy.'

'They are more of a steward. They track wildlife, collect data, check herds, look for signs on carcasses for what might have predated cattle. They help people's interactions with grizzly bears, which are coming here in increasing numbers now that the whitebark pine is disappearing. They are having to switch their feeding habits and come here for the caraway roots. People are finding out about that, so we have more and more visitors gathering to watch them dig away for the caraway, which isn't always a good thing – people and grizzlies don't always mix well, and we don't have park rangers here to supervise the situation.'

'Hmm.' I agree. I have seen people get way too close for comfort.

'Mama, I have a dead fly. He is my new pet.'

Hazel proudly shows us the dead fly on the end of her small finger.

'Isn't that lovely?'

'Yes, he is my new pet.'

'So you said.' We smile at her, and she potters off to train her dead fly.

'The other thing that range riders do is help mimic the wild instincts in cattle – kind of rekindling the bison behaviour, bunching them together. The vulnerable one is the one alone, especially with a calf. When we talk to ranchers about it, we aren't saying what we want is to keep wolves alive, we are saying we want to keep cattle alive. The human presence on the landscape, just with a range rider, makes it more difficult for wolves to come in close. We have our space, and we respect theirs and allow them to use it.'

'Do they get close, though? Do you ever worry about them getting close to the house? Especially with the girls running about?'

'They do get close. One night in January, it was -29°C and a full moon. I was walking the dog pack up the road, and at that time we had five dogs. I got to the top of the hill and turned around. And you know how you are always keeping your eyes out when you are walking the dogs, just so that they are in your periphery? I sensed something wasn't quite right, so I started counting. I got to nine.'

'Oh my gosh, you had been joined by the wolves?'

'Yes!' Malou smiles and shakes her head. 'I wasn't really worried for me, more the dogs, and I did exactly what I shouldn't have done and started running. There was lots of snarling and noise, and it was a very freaky moment. I got home with all five dogs, and the wolves tailed off, but a couple of dogs had wolf saliva all over them. Another day I remember an alpha wolf chased one of them right through the yard till I stopped him with my shovel. So yes, sometimes they do come close.

'I've seen wolves at a distance while I've been range riding, but just that human being there usually helps with the balance on the landscape. It is their territory, but us just being there without needing to be aggressive evens out the playing field a bit. It just encourages predators to move on. It's mainly a dawn and dusk job. The point of dusk is that the range rider will bring the cattle together before they lay down and sleep. You kind of bring in the angles of the herd and wait. It's called settling, and they'll just lie down.'

'So is there any difference since the range riding started?'

'We are working on the stats for range riding right now. There is, for sure, an economic value for the carcasses. In 2016 we found $11,800 worth of carcasses for one herd in the upper basin. They would not have been found otherwise, and there would've been substantial losses. A carcass has to be found within twenty-four hours for a rancher to make a claim. We also verify whether it is a predator kill or not. The range riders are also with the cattle so much that they can check for sickness or injury because those weak animals are the potential predator targets. If they are part of this programme, though, ranchers have to agree that they won't shoot predators.'

Annaliese and her new little black and white calf are watching us through the fence.

'Is Napolean hungry, Mama?'

'No, remember you gave him his milk this morning.'

'This is a great life for them,' I observe.

Malou sighs and leans back, enjoying the warmth of the spring sunshine on her face after the long winter at the top of the world.

'It really is. And they learn so much every day just by being in this place. It's not all just about being pro-wolf, although that is part of it, but people make the mistake of seeing it as some dreamy romantic vision – especially because I am a woman in a mainly male-dominated world.

But we are third-generation here, and the girls will be fourth. We have had more slaughter of animals by predators than anyone I know – and it is awful to lose an animal, you feel violated, like something has been stolen from you – but like it or not, we are right on the interface of it. We are a part of the Greater Yellowstone ecosystem, and this is what it means. I would rather be respected for being open and tolerant than being bitter and angry.

'When I was a girl here, I had a clear vision; our family did. My first real job here, when I was eleven, was as a shepherd to protect our sheep from coyotes. I had a horse that was my own, Jasmine, and we had a great relationship, and I felt very much a part of the landscape. That time was a learning experience for me. That was when I developed my own ultimate experience with the land, my own relationship. We didn't have wolves then, but I saw coyotes slaughter lambs. But I also knew how many I was protecting by just being there. It wasn't about me dominating – I was eleven – it was about being a part of it. It is our responsibility, as ranchers next to Yellowstone, to find solutions that work and support it, including wildlife corridors.

'I know people moan that they reintroduced a "different species", that these were non-native wolves that they reintroduced, but a wolf is a wolf. One of the problems is that, over the years, we have bred our cattle to be more domestic, which has turned them into a dull animal. Yes, they produce good beef, but we have bred out characterisations like curiosity, flightiness, awareness, and that kind of stuff keeps them safe.'

'Annaliese still has a degree of it,' I say, and we both laugh.

'We've started working with White Park cattle in the basin. They come from England originally. They are an ancient breed, and so they are more aggressive, their behaviour is more natural, more 'wild suited'. They bunch together if they are threatened. They will instinctively form a circle with the calves in the middle, and they are good at fending off predators. They are good for beef and a rare breed now, so it will be interesting to see how they do in the next few years. This is one of the greatest conservation challenges of our time – how to manage people and wildlife together on the landscape. And by that, I mean predators too, not just elk and deer for hunters.'

When it's time for me to go, I'm sad to leave for the long drive through Yellowstone. To describe this place as magical might be trite, but I struggle to find the words that sum up the feeling of it. I sit with the view one last time, and I realise the right word is 'peace'. It is that same peace that all us humans need, that we are all exposed to when we are humbled by the beauty of nature, of a system bigger than ourselves. Thomas Moran somehow managed to capture it in his paintings. Some take it another step and believe that the 'bigger than ourselves' bit is God, some believe he created it all for us. I don't, but I do believe nature is bigger than us and – because of our own history of connecting with the landscape and living as part of it – our instincts still kick in when we connect with it. Somehow it just feels right. People might be coming to the wilderness with different ideas of why it is there, but perhaps we all get the same feeling. When I get out of my head and just go with that experience, I realise it is instinctive rather than something that needs to be thought through. We are just part of it.

Clouds form over the mountains again, and in the distance, I can see rain falling over the plains. In Yellowstone, as I climb higher, those spring showers will turn into snow. I feel the sudden need to be with my own pack.

CHAPTER TWENTY-ONE

Feeding apples to wolves

Today I am beyond excited – I'm hoping to meet some grey wolves in person. Although when I look around, I'm perfectly aware that this isn't exactly wolf country. I drive forested roads alongside sparkling lakes and reservoirs. I know it seems like a wolf would be happy here, but I'm no longer in the Wild West, I'm in the very civilised Westchester, New York. It's like a fairy tale world, its mists and manicure completely at odds with the grime and glamour of New York City, which I only left just over an hour ago.

Arriving in South Salem, I pass large (sometimes eye-wateringly so) clapboard homes. Keith Richards has a home here, Ralph Lauren's hides behind stylish gates, Martha Stewart owns the equivalent of a manor house, and Richard Gere even has a yoga studio and restaurant here. Why do I want to giggle when the thought pops into my head of him doing the downward dog – or, worse, the crow?

Each house I pass is different, each charming in its own way. Painted in muted greens and greys and even the odd red one – they seem homely, impeccably decorated, with not a bike, trike or abandoned skateboard in sight. Being a fan of TV's *Location, Location, Location*, I find it hard to keep my eyes on the road.

Here and there, delineating the boundaries, lie stone walls – the kind you find spreading across the dales of Yorkshire. On the east coast of the US, there is a lot that reminds me of home. Many of the place names are British – a nod towards our mutual links, our special relationship.

Martha Handler is president of the board at the Wolf Conservation Center – my destination. After driving for what feels like miles (and that was just her driveway), I finally meet her at her super-impressive and inviting home. She is warm and welcoming, has already asked me to stay even though I am a perfect stranger, and she is just as excited as I am to get me to the Wolf Conservation Center and show me what they do. So, within minutes of my arrival, we shift to her car to travel the short distance to meet the wolves. As she drives, I comment on how odd it is that the Wolf Conservation Center should be here in Westchester – not a place you would expect to find wolves or that seems to be associated with wolves in any way. She laughs and tells me that was her reaction.

'When we first moved here,' she says, 'we rented a home that happened to have a tennis court. Every morning I would get up early to walk the dogs around the yard, and every morning I was surprised to find bones all over the tennis court. Not just small ones either,' she laughs. 'They were big, like deer-sized. I just couldn't understand it. Then one night I was sure I heard wolves howling, and I wondered if we had wolves in the yard at night, but I hadn't heard of there being wild wolves in Westchester. So I decided to investigate, and I told the dogs we were going on an expedition. We found an old trail near the back of the yard, so we followed it and ended up at the WCC. Suddenly it all made sense!'

'Hold on, something still doesn't add up for me. How did the bones get into your tennis court? Were the wolves getting out and coming into your garden at night?'

'No! It was the crows, ravens and vultures. They were picking up the bones from the wolf enclosures as soon as the wolves were done with them, and by the time they got to our house the weight of them must have been too much, so they just dropped them on our tennis court!'

That was the start of something for Martha – already an environmentalist with a passion for wolves. Such a serendipitous moment could not be ignored, and she joined the board of the WCC in 2001. It doesn't take much time with her to realise just how passionate she is about the work they do.

As we pull in to the discrete entrance and up the winding driveway, all is revealed. To be honest, it's a bit of an anti-climax – the place isn't as large or impressive as I had imagined. Martha understands and quickly explains that there are plans for a new visitor centre, but everything takes time and money, and so far, the focus and priority have been conservation and education.

The mission, in this few acres, isn't all that different to that in the vast wilderness of Yellowstone. It's just more focused: to give people in a densely populated state the chance to bridge the gap between the human world and the wild world, the chance to understand and care about the importance and value of wolves. Education, education, education. There are many arms to the WCC's work, but all are centred around that common theme. They work with legislators, they campaign, they are involved in wolf recovery programmes, they offer experiences with wolves to groups of local schoolchildren or anyone who would like to know more. Who could resist an evening camping next to the wolves and hearing them howl through the night? At the same time as getting close, visitors learn about wolf behaviour, ecology, range, and how they coexist with us.

We're quickly joined by Maggie Howell, executive director, who is wrapped in the obligatory green fleece of the conservation worker, enthused and ready to show us around her pride and joy. She is carrying a bag of chopped apples, which I think is a bit odd, but judging by her glowing skin, she probably eats a shedload of fruit.

The sun is out, but it doesn't have the strength to dry the mud on the ground, so we slip a little as we climb the steep slope towards the wolf enclosures. Frankly, I can't get there quickly enough. I have butterflies at the thought of finally getting really close to grey wolves.

'These two,' Maggie waves at two large enclosures to our right, 'we will keep our distance from. Hopefully,' she says with a wince, 'these wolves might be released one day, so we keep human contact to a minimum.'

Alongside the education efforts here run two successful captive breeding programmes for the endangered Mexican and red wolf species. The first enclosure, the one we are nearest to, contains Mexican wolves. I already know that, despite being on the endangered species list for forty years, they remain one of the most endangered mammals in North America. We keep our distance, but we can just see one, dozing peacefully under a tree in the sunshine with not a clue what his subspecies are up against.

Maggie explains that the Mexican wolf was almost extinct in the wild in the mid-1980s after a wave of government-sponsored killing. The only known five remaining wild wolves were captured and used to start a breeding population. Luckily, one of the females turned out to be pregnant, which brought the number up to seven. In 1988, eleven wolves born of that programme were released into the Blue Range Wilderness. With such a small 'starter' population there was a risk that the genetic variation was low, which can open the wolves up to the risk of disease and inbreeding in the long term, but hopes were high – especially because the recovery area set aside was over nine times the size of Yellowstone National Park, where reintroduced wolves have flourished.

Maggie's frustration, and the reason for her wince, is apparent as she gives me the facts and figures – the hard work to breed wolves that would bolster the wild population has paid off. Working collaboratively with other conservation teams to ensure as much genetic variety in the population as possible, they have so far bred many Mexican

wolves here, all of which could be returned to the wild.
Their numbers in the wild still remain critically low:
estimated, at the end of 2017, at only 114 wild individuals
roaming south-eastern Arizona and south-western New
Mexico. These wild Mexican wolves only exist because of
the efforts of captive breeding programmes. Four wolves
born at the WCC have been reintroduced to the wild.
So why is recovery stalling?

As with other species, the Mexican wolf is, not surprisingly,
caught in the crossfire between politicians. The US Fish and
Wildlife Service are responsible for implementing their
species survival programme but are also responsible for
predator control. If a wolf is causing problems with cattle,
for example, it is up to them to remove it and compensate
the rancher for any loss. Then there are biologists and
conservationists who hope to have meaningful input into
the programme, and ranchers and hunters who on the
whole are resistant to the reintroduction as they fear losing
cattle and deer. To be fair, those fears aren't unfounded; the
number of cattle killed by wolves is on the increase – at least
66 in 2018, all of which must be compensated for.

So the wild Mexican wolf population is up against the
human population, and the situation isn't getting any
better – 2018 saw the highest number of deaths, with
seventeen wolves killed. Killing a wild Mexican wolf can
result in jail time and a fine of up to $50,000, but that relies
on the killer being caught and there being enough evidence
for the US Fish and Wildlife Service to prosecute them. In
2015 a man who clubbed a wolf to death with a spade was
prosecuted, and in 2017 an Arizona man was found guilty of
shooting a wolf, but clearly, a couple of prosecutions doesn't
begin to compare with the list of wolves illegally killed.

As with many conservation programmes, and particularly
those involving wolves, rescuing a predator species and
returning it to its rightful place in the wild has turned into

a complicated human scenario. The future of the Mexican wolf on our planet is being thrown back and forth with as much ferocity and passion as a ball in a game of American football. Yet, with such a small wild population, there is a need for urgency to ensure their future. At this rate, I wonder if the various 'teams' involved might have lost the ball altogether by half-time.

In 2017, after much pressure by environmental groups to address the issues, the US Fish and Wildlife Service finalised a recovery plan. According to conservationists, it doesn't give anywhere near the amount of protected land needed and keeps the two wild populations of wolves separate when actually they need protected 'corridors' to be able to move between populations and avoid inbreeding. According to the current rules, if the wolves try to create a territory outside the invisible 'borders' that the government have decided on, even if they aren't causing a problem, they will be captured and moved.

Alongside all this are new rules that make the reintro-duction of captive-bred wolves significantly more difficult. So the adult wolves here in captivity have little hope at the moment of ever being released.

'The government – represented by the US Fish and Wildlife Service – are not making it easy to do adult releases,' says Maggie.

'Why?' I ask.

'They stopped it in 2008 when they had to remove animals from the wild due to complications with cattle. In 2012 or '13, a court order was put in place that meant that the US Fish and Wildlife took the work over. Until then, New Mexico was great, but after that, they pulled out and issued restraining orders preventing US Fish and Wildlife from releasing adult wolves. A lot of people don't like the federal government making local decisions, and now they wanted to release wolves into the area, which the ranchers and farmers do not want. So it turned into a battle between state and government. Ever since then, wild release is not even talked about – New Mexico will not give permission. That was when the US Fish and Wildlife forged ahead with the cross-fostering plan.'

Cross-fostering is complicated. On paper, it looks good, and it has been used with some success with the endangered red wolf population. The idea is spawned from the intention to get as much genetic variation as possible into the wild population and to provide new wolves that are as 'wild' as possible. It relies on monitoring a female wolf in the wild to know when she dens and gives birth. In the rugged, mountainous habitat of the Mexican wolf, that is difficult enough in practical terms, but this plan also relies on a captive female giving birth at almost exactly the same time. Within days, some of the captive wolf litter are taken from their mother, flown to the wild wolf den and added to the wild litter. Bear in mind, also, that the captive wolf could be anywhere in the country, speed is of the essence.

Food is supplemented to the wild wolf to allow for the extra mouths to feed, and ideally, genetic diversity within the litter is increased. In theory, the wild wolf will accept the new pups and bring them up as her own. The captive wolf pups are so young that they have never really 'experienced' captivity and so will be completely wild. If – and it is a big if – they get to two years old, they have arrived at mating age, and only then will they 'deliver' their new genes into the limited gene pool of the wild population. So only at that point is the project really a success. Clearly, there are a lot of ifs and buts.

A revised 2019 proposal doesn't offer any changes; US Fish and Wildlife are still relying on cross-fostering. I ask Maggie what she thinks of that.

'To be honest,' she says, 'it is really unclear as to how much it will help the population genetically. Cross-fostering

is interesting and a good tool, but it is unfortunate that it is
the only tool and that these animals won't have any impact
until two years after the event. It is so difficult to administer.
Last year they aimed to introduce twelve pups but only
introduced eight, and there is no way to compensate for
that missed number. They are aiming for twelve once more
this year, but what happens if they don't achieve that
number again?

'To me, what seems like it would work better is releasing
well-bonded families, animals in relationships, well-bonded
pairs where the female is pregnant. US Fish and Wildlife
now seem to be focusing less on real genetic change and
more on public and state tolerance. They're planning where
to get the wolves from for cross-fostering and forging
ahead with that. So,' she sighs, 'we also have an artificial
insemination programme that will help us predict exactly
when a captive wolf will give birth. We are working with a
reproductive specialist from Cornell University. We have to
catch the female multiple times, and I worry about the
effect, the trauma on the wolf. We can't quantify that.'

'How does it work then?'

'The first step is to catch the selected female for the
programme. We use their natural instinct to do that, so
we have a box with an entrance hole, which we place in one
corner of the enclosure, then a team of us sweep the area.
Effectively, we are herding her. The usual reaction of the
wolf is to go and find safety by hiding in the box. But some
are freaked out, and it doesn't always go so smoothly. We had
one wolf get herself into such a state that she bit her own
tongue. It was horrible.

'The first time we catch the female, we insert a small
implant into her thigh, kind of like a contraceptive implant
but doing the opposite thing – this implant will induce
ovulation. Then we regularly have to catch her to take blood
and check her hormone levels, and when we know that

those levels are just right, we have to catch her to inseminate her with frozen semen. All the "relationships" have been assessed for ultimate genetic variation, so which semen goes to which wolf is carefully planned. The good thing is that we know the due date, so we know if a litter will be available for a cross-fostering opportunity ahead of time and we can plan for that. But even then, you have no guarantee that the pups will be available. It depends on how large the litter is – if she only has two pups, we aren't going to take them.

'And of course they have to be healthy, and there are other factors that we can't even predict. For example, last year, the female we had inseminated decided to dig her own den rather than use the artificial one. To get to the pups, we would have had to dig them out, and we decided we wouldn't do that, it would be too traumatic. Luckily US Fish and Wildlife agreed. It turned out that she had actually had nine pups so she might have been grateful for some of them to be removed,' Maggie laughs. 'One was so little and is still the smallest Mexican wolf we have ever known. He is a year old now, and he only weighs thirty-three pounds. The rest of his litter are like sixty to seventy pounds. He looks like an apple-headed chihuahua living among wolves.'

We all laugh at that. Then Maggie frowns again, that crease between her eyebrows making her smiley face so much more serious and showing just how much she considers the wolves in her care.

'My problem is that we just don't know the effect on the wolves of being caught multiple times. I'm sure it takes its toll. Also, we have to separate the males and females to make sure they don't mate, and they all react differently. I worry sometimes that all this will make them more habituated. Emotionally we have no idea, but all the handling...'

The small wolf curled on the bed of leaves in the sunshine lifts his head to look at a bird that has come a little too close. It may be cold, but he is basking, living in the moment.

'Because education is our priority,' Maggie says, changing the subject, 'we want to share the wolf experience with as many people as possible, to garner more interest and understanding. Clearly, the Mexican wolves themselves have to be kept at a distance from humans, but that doesn't mean that humans have to be kept at a distance from them.' She points to a camera on a tree. 'Thanks to the webcams, we can "tune in" to their world at any time, and if we are lucky, we can share some of their most intimate moments.'

In February 2019, I saw pictures of the successful 'copulatory tie' between Trumpet and Lighthawk as they were broadcast live on the internet via webcams. It's hugely tempting to be anthropomorphic – I would feel distinctly uncomfortable if my 'copulatory tie' were being broadcast on a webcam in the name of education, and one look at their faces might suggest they were embarrassed to be caught in the act. However, they have no idea what a webcam is, and I suspect wolves know nothing of embarrassment or shame – particularly when it comes to having sex outside in public.

'These cameras,' continues Maggie, 'have become a social media phenomenon. Where else would we get to see this kind of wolf behaviour, let alone become involved in the story of these individuals' lives? Did the mating work? Is she pregnant? When will she have pups? Will we get to watch?'

These wolves are in their very own Truman Show, but I shouldn't think they care a jot. Once again, I find myself thinking back to Yellowstone and how part of the fascination is the soap opera of their lives. In 2016 over 300,000 people tuned in to the WCC's cameras to watch Belle (aka F1226) give birth.

'Belle,' says Maggie, 'is a pretty voluptuous wolf by nature, which was unusual as she didn't overeat, but it meant we couldn't truly be sure that she was pregnant. It was May and really hot, over 32°C. We were busy with an event that we had going on in the building, but a viewer noticed her go into the den on the webcam and start pacing a lot, so they let us know. At first, we weren't sure if she was just hot. Then she started digging around and prepping the den, and so we were pretty sure that it wasn't just the heat and something else was going on, she was actually going into labour.

'This was her first litter, so we were really excited. It was also her mate's first litter, and he was watching her very closely. He began to show signs of agitation himself and started pacing up and down the enclosure outside the den. They had been fed with a deer carcass that day, which was still there and, after a time, he brought the deer's head to her in the den. Most people would think that was gross, but we thought it was really cute because he was obviously wondering what he could do to help and thought she might be hungry, so in one way it was actually quite a romantic gift.

'It turned out to be quite a night and pretty traumatic for all of us. We had so many people watching that at some point in the evening the website crashed, but that wasn't our biggest problem. Belle's labour went on and on, and we were worried that she was struggling. Bear in mind that we will never intervene unless it is life-saving because we want to keep human contact to a minimum, but we continued to worry, so we got the vets over to watch. After watching her for a while, they said that if this were a dog, they would be intervening, so we made the difficult decision to go in. We had to catch her in the den and give her subcutaneous fluids and calcium, which I learned that night can help induce the labour.

'When we let her go, she didn't go back to the den, which was understandable because we had taken her out of there. So instead she hid somewhere in the enclosure, and because it was dark, we had no idea where she was. That was the worst bit for me because we had no idea what was going on for her. By now, the website was back up, so I had to spend time answering questions about what was going on, which helped me keep calm – but of course, I was really worried.

'Finally, around midnight, we were standing quietly beside the enclosure, and we heard small sounds – squeaking and grunts – and so we were pretty sure everything was OK. At first light, we could see her in the brush. She was flat out, exhausted. Three healthy pups were nursing, and so we were all relieved, and then we noticed that she was using the deer head from her mate as a pillow. He must have brought it to her again in the night!'

Happily, mum and all three healthy pups survived, and the new wolves were named after the three volunteer vets who saved them.

Digital wolf watchers access the lives of their favourites from all over the world, and it sounds like they are as loyal as the real ones I have met. They get involved by letting the centre know what has been seen and adding to the information gathered. Just like the wolf watchers in Yellowstone, here wolf conservation is not just down to the scientists – any observation is valuable, and encouraging that passion for wolves around the world is just what these guys want to do.

I'm not here to experience anything remotely, though. I'm here to get close to wolves – and in particular the grey wolves, the same as the ones I have been watching in the wild. As we walk towards their enclosure, they walk towards us, interested. Three wolves – two white and one silver-grey. It is impossible to overstate their presence when you are this close. There is something so familiar about them because in one way they seem like giant dogs, but there is also something more. I am captivated right away.

'Philippa,' says Martha, 'meet Zephyr, Nakai and Alawa.'

My first question is always the same: 'Why can't they be in the wild?'

'They were bred privately with no place for release,' Martha replies, 'and they were never even part of any release programme. Also, they were too socialised with humans to be released. So their home is here, and we use them as ambassador wolves. They are, although they don't know it, doing their bit for conservation. Their value is an important part of the centre, they give people the chance to connect, to really have a tangible experience with a wolf, which most people will never get because most people won't get the chance to see a wolf in the wild.'

Maggie offers me a chunk of apple. The wolves go crazy when they see it.

Zephyr is easily the most handsome. He has a dark grey, almost black saddle, tail, ears and muzzle, while the rest of him is silver. He is seven years old and paces excitedly, winding around his dainty sister Alawa and younger brother Nakai, who are both white.

I realise the apples are not for our health after all.

'Throw it over the fence.'

I'm surprised. I didn't really have wolves down as apple fans, but I do it anyway, and the reaction is hilarious. All three leap up into the air for the healthy snack. I am feeding apples to wolves. Maggie and Martha each grab a handful too, and we throw apples over the fence and laugh. The huge wolves jump high, each trying to intercept the treat before the other. Zephyr is the highest by far, showing us his silver underbelly as he twists and turns, getting at least seven feet into the air but always landing with utter precision. He can seemingly throw himself at any angle to catch a piece of apple. Maggie explains that he is the 'leader' of this small pack.

'Well, he thinks he is, but his little sister is pretty good at getting her own way too. She just does it less obviously.'

She isn't able to jump as high as Zephyr, but I notice a steely look of determination in her orange eyes.

When all the apples are finally gone, I stand close beside the fence. One by one, the wolves come to say hi. Zephyr is first. He sidles up to the fence and rubs himself along it. I crouch down, and he sniffs at my face looking deep into my eyes. Our noses are almost touching, and I can smell him – a deep, earthy dog smell. His dark eyes are ringed with black. His pink tongue comes out and licks his nose, and he moves slowly away.

This is not a typical pack but more of a family unit, so the dynamics are different. Dominance often shifts and their personalities are apparent. Zephyr is cautious but inquisitive, so he is usually the first to check things out. Nakai is the most submissive. He is still only four years old and very playful, and he is the next one who comes to see me, rubbing the side of his chin along the fence level with my face, his eyes closed as if he were a cat. He presses so hard his fur squeezes through the gaps, and it is very difficult to resist the urge to touch him, but that remains out of bounds.

'He's a bit of a goof,' says Maggie. 'Really they have structured themselves about what is best for the family here.'

After a few paces, he turns and does the same thing again on the other side of his body. There is no denying the greeting, no denying the fact that we both wish this darned fence weren't in the way.

Finally, Atka (whose name, says Maggie, means Sweet Pea) approaches. She seems so demure, taking her turn, her yellow eyes in her white face showing no sign of threat or fear. She takes her time, sniffing at my face. Then the wind blows, and some of my hair goes through the fence and into the enclosure. She licks it, and I can feel the warmth of her breath. I am blissing out.

Then it starts to get a little competitive. Zephyr and Nakai have decided they want another go at greeting this new person at the same time, so there is a little snarling. Zephyr raises his lips to show his teeth, not two feet away from my face, and now I am grateful for the fence. Nakai backs down immediately, lowering himself with his head facing up in a submissive gesture, but he does bare his teeth, and now his yellow eyes, lined with black, look fierce. Yet he still stays as close to me as he can. Zephyr is right on top of him, straddling him. Whether it's interesting visitors or apples, there is no doubt who is in charge today.

'Not everyone gets greeted like that,' says Martha. 'They steer away from some people altogether.'

I am, rather stupidly, beyond proud that I might have some kind of affinity with wolves. After saying hi, the wolves quickly lose interest in me, however, and I fully appreciate that it could be because the apples have clearly run out. Maggie, Martha and I all settle on a nearby bench to chat, cold but reluctant to leave the wolves altogether.

Suddenly Maggie takes a deep breath and emits a very odd-sounding, high-pitched squeak. She does it again, and Martha starts too but in a deeper tone. Within a second, Zephyr lifts his chin to the sky, assumes an expression of deep bliss, closes his eyes and begins a mournful chorus. Wolves in big enclosures all around us join him, and we are surrounded by tuneful, soulful howls.

When he lowers his chin, signalling the end of the session, there is no sense of embarrassment at his 'automatic' howl reaction, just a sense of 'this is what we do'. As if it were no different to drinking a cup of tea or having a chat over the garden fence.

And yet it is different. It doesn't matter how many times I hear wolves howl, there is something very odd about it, as though it is just the first taste of a world we will never truly understand.

CHAPTER TWENTY-TWO

In memoriam

Even though there are only a few volunteers here today, the place is buzzing in preparation for a big benefit that they are hoping will raise significant and much-needed funds. Silent auction prizes take up any spare perch in the building: crates of wine, framed paintings of wolves and horses, and a picture that really captivates me. It shows a white wolf with wings, clearly headed to heaven, while on the ground, a young girl holding a white pup reaches out her hand as if she were trying to pull him back. Martha catches me looking at it.

'This artist is amazing. We are actually flying her in to speak at the event. She is just fourteen. Her name is Bria, and she sells all her artwork to raise money for endangered species. She is a huge fan of the webcams and has been following our work for years, so we are lucky to have her donate a painting to us. One of her paintings made us $15,000 last year at an auction. So far, she has raised over $35,000. She has her own Facebook page and has donated to the Jane Goodall Institute, the International Fund for Animal Welfare and us. She's even published colouring books.'

Everyone is discussing the gala, and it all seems to be in hand – there's nothing for me to help with – so I check out Bria on Facebook. Another digital supporter, passionate about wolves, she started facesoftheendangered.com when she was just eight years old and uses the hashtag #kidscanchangetheworld. Her work is colourful – childlike, yes, but with a sophistication nonetheless, and it is precisely that childlike quality that pervades it with an optimistic, imaginative theme.

In the middle of the bustle, with her arms full of wine bottles for the auction, Maggie stops to talk.

'That white wolf in the painting is Atka, our superstar. We just lost him this September, and I can only now just talk about it without crying. He was an ambassador wolf but a very special one, an arctic grey wolf who everyone adored. He had thick white fur and the most beautiful nature. He came to us when he was just eight days old.'

'What made him so special?' I ask.

'His personality. He was so approachable, so friendly. Nothing phased him, he was able to do his own campaigning,' Maggie smiles. 'We took Atka to DC when we were campaigning for the Protect America's Wildlife Act. He was tangible – people had never seen a wolf and so seeing him right outside Congress made a huge impact. He had no fear. He even went to the bathroom in Congress, making another impact – he was clearly marking his territory. Everybody loved him so much. I will be speaking about him at the gala tonight, I'm not sure how I will do it without crying.'

Later the same evening, I find myself at Le Chateau. When you come from Europe and hear 'Le Chateau', you expect turrets, grey stone, a fairy tale castle with a bit of box hedging and topiary. This is not that kind of chateau. It is a hotel that to an English person looks not the least bit historic or French. However, inside it is exquisite. There is more wood panelling than I have seen in a long time, and the staff are lovely.

We are all dressed in our finery to celebrate nineteen years of success in the WCC's mission of education and conservation. This is a far cry from the anything-to-keep-warm wardrobe of the conservationist that I have become used to, but it turns out there is as much passion for wolves in Salem as there is in Yellowstone – it is just expressed in a different way.

At the appointed hour, two hundred glamorous guests pour into the chateau. They have spent $125 each on tickets, so the WCC team are keen to make the evening a raging success. As they arrive, they pause for photos, posing with daft accessories or holding cardboard cut-outs of the wolf ambassadors I met earlier. Little black dresses are everywhere, and expensive jewellery sparkles on earlobes. Perfectly manicured hands reach for the welcome drink – a wolf-inspired cocktail made from a mezcal called Montelobos (meaning 'mountain of wolves') that, given the rather luxurious nature of the event, is made from organic agave.

Cocktail in hand, the guests are invited to peruse the items in the silent auction. In no time at all, they are bidding and outbidding each other for the various prizes on offer. Looking around, I imagine most of the people here are pretty wealthy, but there is no doubt – at $125 a ticket – all of them care about wolves, and they are gearing up for a fun evening. Just how much they care becomes apparent when, after a few more cocktails and some rather lovely appetisers, the speeches begin.

The evening is dedicated to Atka. I had no idea just how much local people were attached to him. Although these are plush, civilised and warm surroundings compared to the wilds of Yellowstone and Montana, the emotions are the same, and I am reminded of the strength of the grief that followed Spitfire's death in Silver Gate.

Maggie gets up to speak. Her voice does indeed crack, she holds her hand to her chest and struggles to keep the tears at bay, but of course they come as she talks about what losing Atka meant.

'He lived to the great age of sixteen,' she tells us, 'and died peacefully in his sleep with his "family" around him.'

His pack were humans but, judging by the atmosphere in the room, they were no less loyal than a wolf pack. I am reminded of the wild wolves in Yellowstone who rarely make it to ten years old.

We watch a video about Atka, whose name means 'guardian spirit'. He was stunning: pure white with just a shadow of darkness beneath the thick fur. His eyes were

pale green, lined with black matching his black nose and
mouth – he was pretty much the perfect specimen of an
arctic wolf. The video shows him as a small, grey puppy, then
as an adult on the beach, on stage, in front of Congress.
In one image he is surrounded by adoring fans taking his
photo. We watch him howling in snow, and rolling in sunlit
autumn leaves. For any wolf, his was indeed an extraordinary
life. His howls echo around the ballroom and, as the white
face on the screen slowly fades to black, I look around me –
among the beautiful people of Westchester, many faces
glisten with tears. One husband massages his wife's shoulders
as she silently cries. There is not a whiff of stiff upper lip
here. I have just been introduced to another legendary wolf
who touched many hearts.

Next, Nico stands up to speak. He is only ten. Nerves and
an audience tend to make children talk even faster than they
usually do, as if they just need to get it done but, that aside,
this kid spoke from the heart.

'All I know is that Atka was like family to me. His spirit
was like family and will stay in the centre where he belongs.'

Atka clearly lived up to his name.

This evening, under the glassy chandeliers, sipping on my
'wolf' cocktail, I notice the association between spirituality
and the wolves that so many have is really apparent. The work
of the WCC, of course, is intended to give an appreciation
of the science and ecology of wolves, but tonight people are
just as focused on the mystery. It's clearly not a new thing –
Native American stories talk of the strength of the wolf
spirit. The artwork displayed for the silent auction contains
many depictions of wolves with a spiritual connotation.

Ten-year-old Nico's speech finishes: 'When the kids ask
"Where is Atka?" I will tell them that he is still here, we just
can't see him anymore. Atka worked to create a better world
for wolves and so will I.'

There is a resounding round of applause – enough to
make a glass chandelier tremor. The tears are dried for now.

Very quickly, the tone of the evening changes. In fact, when the auction begins, it is fast and funny, and everyone is so inspired by Atka's story that they dig deep. The total tally for the evening is $110,000 – a massive $60,000 more than last year. All that money will go to education and conservation.

Even though most people in the ballroom tonight will never see a wild one, there is no doubt of the extent of the passion for wolves here in rural New York state.

CHAPTER TWENTY-THREE

I love wolves and people

'The Lower Gros Ventre pack were all removed because of conflict with cattle in 2012.'

Ken Mills is matter-of-fact about this. His bright blue eyes above a tawny beard are fixed on the road. They have to be – he is driving up the steep and winding pass that leads out of the valley and into Idaho. The rain that has soaked the valley all morning has now turned to snow beating against the windscreen as his Wyoming Game and Fish truck climbs to altitude.

Ken has allowed me to come with him to check on the Chagrin River pack. He thinks he knows where they are denning and wants to find out for sure. To do that, he is going to set up some camera traps and leave them for a few weeks. He did warn me that he might not be the best person to meet up with because he 'isn't really a talker'.

'Why are they called the Chagrin River pack?' I ask. 'Where is that? I didn't know we had a Chagrin River.'

'We don't. A guy first discovered them in 2010 when a couple of wolves showed up, so we said that he could name them, and he chose to name them after a river in Ohio!'

Ken is the man in the middle, the man who has to hear all the demands from hunters, ranchers, government and wolf advocates. It is his job to try to keep wolves and people happily living together. He is the wolf biologist for Wyoming Game and Fish, and he works in the large carnivore section alongside a bear biologist and a mountain lion biologist. He is responsible for overseeing the monitoring of all the wolves in Wyoming, and he proposes management decisions based on what he sees.

'I know wolves,' he says. 'I'm not the decision-maker; those are my superiors, but I know what is going on on the ground.'

'Does it actually work, though, just taking wolves out? Because doesn't the evidence suggest that when you do that, other wolves will just move into that space?'

Ken looks in the rear-view mirror as he answers. We are towing a long trailer around these tight, high bends, and on it is a 'side by side' – a small 4x4 that we will need to get where we are headed.

'Every year you manage wolves who are in conflict with cattle, then you tend to drive down cattle depredation, not just immediately but in the following years as well. There has been research done by Liz Brady that shows what the time gap is. What you are doing is removing the ones with the inclination, you are managing an established pack that, in my experience, can be more problematic – partly because nutritional demand has gone up. They have new pups and yearlings to support, so they usually start depredating cattle when the elk calves are no longer around. If you look at pack stressors, if you kill pups mid- to late summer you release that strong nutritional stress, and you see a shift in behaviour at that point. If you remove that pack, then you start over, so you can incur a two-year delay. I'm not saying that we enjoy it; it's just what the data and my own experience indicate.

'It makes sense of what we know about wolf history for individuals. When people know that I'm a wolf biologist they often ask me, "winter's got to be hard on the wolves, right?' and my answer is "no, September is tougher." The pups are at their maximum growth rate, elk calves are bigger and harder to get hold of, and the elk are on the move, so in September the wolves are coping with the highest nutritional demand from the pack overlapping with the least nutritional availability.

'Wolves are really adaptable and resourceful. If they are under pressure and they need to find a different behaviour, then they will, and that means they will switch to eating livestock if they have to. But it isn't their default. I have seen wolves moving through a herd of cattle but not hunting, and the cattle know it too. Not all wolves just see cattle and feel the need to hunt.

'Wolves and prey have an intimate knowledge of each other, I believe. Elk probably know individual wolves in a pack and who they need to worry about. We can't put our "humanness" to the animals that live out there in the wild. Wolves know their neighbours, they know the herds that live beside them, all that feeds into their decision-making and behaviour. They live alongside each other – how can they not know?

'It is an intimacy that we don't experience – what is going on out there in the landscape day after day, night after night. What do we really know when we might only be out there once a month? They know their territory like the back of their hands – they know when they have an opportunity to extend it, for example, if a neighbouring pack are not doing so well. If there is an advantage to be had or out of necessity, they will act aggressively. In 2018, here, we saw the highest incidence of wolves killing wolves that we have ever seen, and we write all that up in an annual report that everyone can read.

'We live in a unique place really, especially because of the wildlife that surrounds us. That is why the large carnivore section was started in the early 2000s – to assist with all of this. It's a big workload here because there is such intensity, we are specifically dealing with conflicts. We manage nothing else like wolves; they are unique, being a social, loving carnivore. Nationwide you have the APHIS [Animal and Plant Health Inspection Service], under the US Department of Agriculture. Part of their role is to manage conflict – and Wildlife Services too – but here it is so intense that we need regional personnel that are specifically assigned to do large carnivore work. When you live in the "real world" with wolves as we do, you usually end up in the middle of it, you don't have a choice.

'I see the rise of hunting now as a popular sport. There are more magazines and videos than ever before, and hunters are going to value elk more and other competitors less. It isn't just a recreational thing that a family might do to fill their freezer any more. There are outfitters out there making a living from it, and elk antlers sell for good money now. Any time we have competition for something we value, we tend to look negatively on that.'

Safely down the other side of the pass and back into the rain, we bump down a remote track and pull up next to a gate. There is a sign that might explain where we are – if I could read it. I can see it says something about private property, but it is so covered in bullet holes where people have been shooting at it that I can't really read anything else. Ken points at the slim, muddy track behind it.

'This is where we change vehicles,' he says. 'That's where we are heading.' But we sit in the car, looking at the raindrops running down the windscreen. 'It might surprise you to know that wolves are not my number-one priority,' he says quietly.

I look at him and see he is smiling, almost as if he is joking.

'What do you mean?'

'It's not that I don't love wolves. I do. I'm passionate about having them on our landscape. But I see people. I care about people. People are my number one – they have to be; otherwise, I couldn't do my job.'

Ken watches a crow who has fluttered down from a nearby tree and is now next to the truck, picking up the worms on the muddy track.

'I wonder why he is going for the dead ones and not the wriggling ones?'

The crow marches to and fro between the puddles, happily feasting while rain drums on the truck roof.

'There is one quote,' says Ken, 'that changed everything for me in my life but also in how I do my job. It's by C.S. Lewis – you know him?'

'Of course,' I laugh. 'I'm British, aren't I?'

Ken laughs too.

'Of course. Well, he wrote, "You have never talked to a mere mortal," and that changes everything for me. When you look at people in that way, when you understand that everyone has a story, it changes the way you are with people. That enables me to do my job, to be in the middle of all this conflict.'

'I get it,' I say. 'Even the person you might overlook has a story that is usually wonderful. They have a good reason for being who they are; you just have to listen to it and try to understand.'

Reluctantly we get out of the car and change into our rain clothes and boots in the muddy lane. Ken climbs onto the trailer and starts up the 4x4. It rolls down the ramp, I clamber in, and we are off heading up the trail.

'This is a horrendous trail,' Ken yells over the noise of the engine.

'Oh, good!'

He laughs.

'There are some really sketchy points, so I'm not sure how far we will get in this mud, and the first one is around this bend.'

He is right – the way is covered in huge rocks that hit the bottom of the vehicle with loud cracking noises that make me wince. Ken drives with courage, not taking his foot off the accelerator, and we power over everything. From time to time we are thrown against the doors as crevices, holes and drainage fissures tilt the 4x4 almost at right angles. It is great fun, and there is an element of letting the terrain do the steering while we just power up it.

'This vehicle is amazing. I can't believe we haven't blown a tyre on these rocks, they are so big.'

'I know. Well, we might have done – it's hard to tell.'

We drop off another crevice.

'You know my favourite tool at the moment?'

'What's that?' Conversation isn't easy. I'm gripping onto the handle in front of me just to stay on my seat, and every time I try to speak the bumps take my words away.

Ken negotiates another slide, and replies: 'A horse.'

I laugh.

'Seriously. A horse can take you through most terrain, especially in spring when there are still patches of snow and mud. A horse isn't noisy, not scary to wildlife. I love doing my job on horseback, it's really practical.'

Finally, we reach some flatter ground and leave the rocks behind us. It's beautiful up here. Newly unfolded leaves, lime-green, flutter against the white trunks of aspen trees surrounding a small lake. The new spring grass is bright like a well-kept lawn, and the now flat track winds gently ahead.

'I would say we are about halfway there now. It is a beautiful spot, eh?' Ken powers forward. 'The grass is short because it is grazed, so we have some cattle conflict in this area sometimes but not at the moment.'

'Do the ranchers here take any anti-wolf measures? Are they interested in that?'

'No, not really.'

'Why not?'

'I think it comes down to trust. The same groups that tell them what to do and how to deal with wolves are the same groups that will support relisting, and that isn't what these guys want ultimately. They'd sooner be able to hunt wolves and keep them away.'

We reach higher ground where there are some snow patches, which our vehicle does not like. For a while, we can skirt around the edges of them, but we eventually come to a point where all the ground is covered in snow. Ken cuts the engine. Our trusted steed can carry us no more.

'This here is where we start walking,' he says, hoisting his rucksack onto his back. 'I reckon we have a mile or so to hike. Let's not forget the bear spray.'

We are above 7,000 feet now, much higher than Jackson. The climb is steep, and we are potholing into the soggy snow and fighting our way through branches. I get breathless pretty quickly, but Ken is still chatting away.

'I think there are three wolves up here who should be denning. They had several den sites last year, and it will be interesting to see whether they use the same ones this year.'

'What decides that?' I pant. 'I mean, sometimes they seem to use the same one year after year.'

'I think tradition dictates some of it, but we don't really know. Sometimes they can skip a den for a year or so and go back to it. That could be because it has mites, maybe – or who knows? Wolves in the trophy areas, where they aren't protected like they are in the parks, operate as you would expect generally. There is a density dependence, and we see rapid growth rates when we are at middle levels for the population. So, in terms of population dynamics, it is pretty predictable: when the population is small we see a slow rate of growth, then a faster rate of growth and then we get to the point of carrying capacity where the population has reached its upper limit and becomes unsustainable.

'We aren't at carrying capacity this year, but we were last year, so we saw more diseases. Mange especially killed off one pack, and we saw more wolves killing wolves and that intra-specific competition kicked in. We had never seen that much before, but in my mind, that was due to unstable management during the last few years, which caused a lot of instability in the population as it was operating as it would naturally. We had too many management approaches when the wolves were delisted and then listed again. When they were delisted, that led to an increase in population size, which then declined again. All that would suggest that in 2012, when Game and

Fish were regulating hunting, there was a decrease in the population plan according to the management plan. In 2013 the population decreased. In 2014 before the hunting season, they went back onto the endangered species list. We predicted that if we had killed forty-two wolves in that season, we would have been at our target of 160 wolves, but there was no season, in the end, so 205 wolves remained, and we had more wolves than in 2011.

'Everything changed when the wolves were on the list. The pressure was released, and that led to a rapid expansion in the population size, and then we ended up with an increase in stock conflict. The number of wolves killed for that went up, so we had 113 killed in 2016 for conflict with stock.'

I'm struggling to keep up – both mentally and physically. Ken knows the population numbers and the history so well, and he knows how to climb a mountain.

'The wolves were delisted again in 2017, and we had a hunting season based on our Game and Fish report. But because no one had really been on the ground, there were twenty-nine more wolves than they thought. So that meant in 2017 we still had well above 200 wolves, which is where we see signs of density dependence kick in.'

'So, what does that look like?'

'When we are looking at density dependence, we have to ask what the stressors are? Wolves are more unpredictable in terms of their behaviour, and we have to look at the disease impact and the competition impact. At carrying capacity, the whole thing is much less predictable. So by 2018, as I said, we had disease, wolves killing wolves, competition, and far fewer pups being born. I know the factors that will kick in but not exactly how they will play out. Our data – the data that I gather every day – is really rich and useful here. This natural population is kind of like a laboratory. Wolves are different from other species, so this data is valuable. If we have strong long-term data, we have a high level of predictive power. Do you need to stop for a moment?'

'I'd love to. How high are we now?'

Ken consults his watch: '7,920 feet. Pretty high,' he says, smiling at my breathlessness. 'Take a look at this tree over here.' He leads me to a large aspen tree in the middle of a stand of silver. The trees up here have no leaves yet, as it is still too cold for such delicacy. Ken points at some marks on the trunk. 'See these? Know what they are?'

I'm thrilled. If I'm right about what I think I am looking at, these marks are what I've been looking for since I arrived in Wyoming..

'Bear scratch marks?' The long dark lines are pretty clear – deep claw marks gouged into the white trunk. I look up. 'But they go so high!' The marks go at least fifteen feet up from the ground.

'He was climbing the tree, for sure.'

We carry on hiking and then both see a scat pile at the same time. It's a big poop!

'Bear?'

Ken pokes at it with a small twig, tearing it apart.

'No, it's wolf. See all the hairs in there? Elk hair.'

'But why is it full of pine needles too?' I'm pretty sure the average wolf doesn't eat pine needles.

'I guess if you throw a steak on the ground and you don't have silverware and you are just eating it with your mouth, then you don't get much of a chance to pick the pine needles off it. And they are like dogs in that they do eat grass at times.'

Ken sniffs at the twig and offers me the chance. I'm used to sniffing otter poo, so I give it a go.

'Can you smell the mustiness?'

'Yes.'

'That is also what distinguishes it from bear. It's kind of richer.'

It is very fresh – we are close to wolves.

Snow starts falling. A little further ahead, Ken bends down to look at some tracks in the mud.

'Wolves. See the four toes?' I nod, and he continues: 'See the pattern of the feet, the way they are almost on top of one another? This wolf was trotting.'

We carry on hiking uphill – potholing through the snow where it is soggy, walking on the top of it where there is a crust – until we come to a trail and then a clearing. The aspens and conifers suddenly give way to a white mountain meadow. Now we can walk side by side, and the going is easier. We no longer have to navigate the low branches, but the snow is falling heavily.

'This is where they had a rendezvous site and a den site last year. It's a great spot where a lot of wildlife trails meet and a place where moose, deer, elk and bears will all come, so I'm keen to get some cameras up here.'

I look eagerly around me, but there are no tracks in the snow.

'The wolves are probably denning now,' says Ken, 'so we are unlikely to see them, but I once came up here and there were ten pups just hanging out and playing. It was the year that a new male had joined the group and that often happens when a new male comes in – he mated with two of the females. There was only one adult with the pups, a young one, clearly left to babysit. She did a good job – when she saw me, she led them all off into the trees to hide. I think that same adult male is still with them. We collared him once, but now he is so old that his radio collar has stopped working. He must be about eight now. That is a good age for a wild wolf around here.'

When we get to the other side of the meadow the snow is so heavy that we take shelter. Our only choice is to stoop under the broad branches of a grand conifer. Under there it feels as though we are in a little cave, and we sit on a fallen log and watch the snow fall 'outside'.

'I've trapped wolves up here to put collars on them. I caught three pups last year.'

'What kind of trap do you use?'

'A leg hold. It's based on the old Newhouse traps, but it doesn't cause them any harm. I check them regularly, so they aren't in there for long, but that collaring is worth it. Out of a population of around 195 we have collars on around ninety-five of them. That is an incredible feat, especially for the amount of clear data – that is where it is really valuable.'

'Can I ask why Wyoming manage to the bare minimum in terms of population? Why not just let the population be?'

'I hate hearing that,' Ken sighs, 'but it is a long and complicated answer if you want to hear it.'

'We haven't got much else to do,' I say, indicating the falling snow.

'OK. Back in the early 2000s, Fish and Wildlife were preparing to delist wolves as they came close to the criteria numbers. This had been on the table as part of the reintroduction plan. Three recovery zones were in operation: central Idaho, north-west Montana and the greater Yellowstone area, which includes Yellowstone, Wyoming, the lower third of Montana and the eastern third of Idaho. We had agreed that we would delist when there were at least 300 wolves with at least thirty breeding pairs equally distributed between those zones for three years in a row. The service actually re-evaluated those criteria so that there would be a buffer, so they actually put the numbers up by 50 percent, so that meant we were looking for 450 wolves in the population with fifteen breeding pairs in each recovery area. Then, in 2002, the boundaries got changed.'

'What do you mean?'

'They changed the boundaries of the recovery zones and brought them back to state lines.'

This was a big challenge because looking at the state boundary, rather than the Greater Yellowstone Area boundary, halved the area of land that wolves needed to be in, yet the figures were never adjusted for that.

'In March 2008 the wolves were delisted, which means they were taken off the endangered species list. At around that time, Idaho had 800 wolves, Montana had 600 wolves and Wyoming had 350, there were 90 with 7 breeding pairs in Yellowstone.

With wolves delisted, management was handed back to the state, and so Wyoming's management plan went into effect.

'So that was really where my work was meant to start,' Ken continued, 'but'.

'I thought I could hear a 'but'.'

'Well, the decision was challenged. The evidence to do that came mainly from a study by Bridgett von Holdt and her team from UCLA on gene flow between the populations. There was concern that in Yellowstone – with seventy to eighty wolves, and only seven or eight breeding pairs outside Yellowstone – there wouldn't be enough genetic interchange. They needed more time to analyse their results to prove that there was adequate gene flow between the GYA and other recovery areas. Until that could be proven, the wolves were back on the endangered list. I arrived in Wyoming and started my job, which was to implement the management plan, on 1 July 2008. And they were relisted on 18 July, so they were right back on the endangered list.

'This was weirdly a case of litigation being faster than science. It was the politicians who said that wolves would have a problem across the northern Rockies, but the data wasn't analysed yet. In 2009 that study broadened across the northern Rockies, and the scientists discovered there was nothing to worry about with gene flow after all, but of course, the wolves were already back on the list.

Big flakes of snow fell steadily around us, filling out our tracks in the mountain meadow. We were so far from anywhere that there was no other noise at all. I wished with all my heart that a wolf would walk slowly by, carving out his own tracks, headed for a den in the trees, but I knew it was unlikely. Whether or not the science backed it up, I couldn't help but feel glad that the wolves had been put back on the list – I still hated the thought of one being shot, of blood spilled onto the snow and a lifeless wolf in front of me.

Wyoming contains two national parks, Grand Teton and part of Yellowstone. Wolves would always be protected in those places whether they were on or off the list, and that was good to know. Outside that area, it would be up to the state to manage them.

Ken continued, his brain full of facts and figures that just poured out.

'In 2009 in Wyoming, 85 percent of the state was designated as predatory zone. In this zone wolves were not protected, they could be hunted and killed without a license and killed if there was any suspicion at all that they might be about to 'bite, wound, grasp or kill any livestock or domesticated animal.' The north-west part of Wyoming (including Jackson Hole) was declared a 'trophy hunting' area where numbers would be managed tightly and consequently, only a certain amount of hunting would be allowed, and that would be done by using licenses or tags. We would only allow a certain amount of hunting.

The US Fish and Wildlife Service wanted to send Wyoming a clear message, "until you have a full trophy area, we are not delisting you." Wyoming challenged that by saying there is nothing in the Endangered Species Act that requires a full state trophy area. And it does actually acknowledge that with any endangered species we have to be aware that there will be areas where there is conflict.

In 2011 wolves were delisted in Montana and Idaho but not Wyoming. They cut Wyoming out of it because of the trophy zone and predator zone, reissued that 2009 rule and said it wouldn't be challengeable in court.'

Kens sighs deeply,

'When Mead was elected, the goal was to get wolves delisted in Wyoming, and negotiations started with Fish and Wildlife to develop an agreement where documents like our management plan would meet their requirements. I was asked to advise on biology. In that agreement, there were ninety wolves and seven breeding pairs in Yellowstone, which meant that we were going to have to contribute more, so we agreed to manage ten breeding pairs and one hundred wolves. To put that into perspective, we agreed to manage for the original amount agreed at reintroduction, but in 25 percent of the suitable recovery area because that original amount was set to the GYA tri-state area, not to just Wyoming. One of our biggest problems is that there just isn't suitable habitat outside that trophy zone – too much comes into conflict with cattle and sheep etc.

'In 2011 wolves were delisted by Congress. Idaho has 800 wolves and only has to maintain 150. Montana, with 600 wolves, only has to maintain a quarter of them. Wyoming has 350, so we would have to maintain half of them, and there are ninety wolves with seven breeding pairs in Yellowstone. Are you still with me?'

'Yes, just about.'

'Because what I'm talking about is proportions and it isn't easy to explain or grasp, but it is the key to the challenges here. So, Wyoming has to keep double the amount of wolves that any other state has to maintain. Wyoming has to pull the slack. The difference is the onus on us versus the other states because of Yellowstone. We are being asked to maintain – outside of Yellowstone – a larger proportion than any of the other states, and we have less suitable habitat because we have a lot of wolves to maintain in conflict areas. I want people to understand because Wyoming is pulling more, proportionally, than any other state yet being criticised for it. Also, in our management plan, we committed to maintaining a population buffer above the minimum to allow for variation and make sure that we were always meeting that minimum. This is something that Idaho and Montana did not do. So with that, we were working toward delisting in 2012.'

'OK, so what was the number? What was the buffer?'

'Ha!' Ken laughs, but he doesn't sound very happy. 'That was the issue. What was the buffer going to be? Well, of course, it wasn't that simple. I said that we will let the data tell us what that needs to be from year to year. We have to go along with the variation from year to year. If we set a number, we will have to change it, and people will think that we are being inconsistent. We have not hunted wolves in Wyoming for years and so with a population unmanaged for so long, numbers were going to fluctuate again till they settled down. So our suggestion was that we didn't agree to a number that would have to stay the same year after year, but we agreed to a concept, which would be guided by data. We had already agreed to a number proportionately greater than any other state, and the fact that we even had a plan for a buffer was different. Guess what?'

'What?'

'We lost precisely because of the buffer – because we didn't put a number on it and people didn't trust us because of that. It was unfair. We don't whine about it, but people give Wyoming zero credit for the commitments we made. Whether or not you like the trophy game area is an ethics call. I get that.'

He pauses, and we watch the snow fall. There is no sound other than that of dripping water and the odd thump of snow as it gets to a critical weight on one of the conifer branches that finally bends and lets it go.

After a while, Ken continues.

'My interest has to be how to manage the numbers, so the wolves are at a safe population. And to do that, I come back to the data every time. It is difficult to make sure you have the right amount of breeding pairs. If you graph the numbers of wolves living in the trophy hunt area and the number of pairs, there is a really strong correlation. More wolves mean more breeding pairs, yes, but if you apply something called an R-squared analysis, then it's very clear that 95 percent of the number of breeding pairs is explained by the number of wolves at the end of the year. So I can look at the data and be confident that when I analyse it I can set the buffer number for the following year, and I can do that with a 95 percent confidence level.

'So if the data says that we manage for 142 wolves in the trophy game area, at the end of the year we can be confident that we will have at least ten breeding pairs, more likely thirteen or fourteen. Actually, when we had 140 wolves, we had twelve or thirteen breeding pairs. So we are managing at 100 percent confidence that we will be above the minimum. Managing for a hundred wolves means ten breeding pairs plus a buffer on top of that. But if we manage for more than 142 wolves, we are allowing for other variations too – the type that might be harder to predict. For example, we all knew that density dependence was going to happen to some degree, but not what the results would be exactly.

'Going into 2012 we had 186 wolves, so there is no way you can say we are managing for the minimum requirement – we had 152 at the end of 2018 because we had to offset what had happened in 2014 to get away from the conflict issues etc.

'I have to have this detailed perspective because it is my job, but people don't see it or sometimes don't know the details. I'm not asking for everyone to love our management plan but to give us some credit because the data is driving this at 95 percent certainty, which means that one out of twenty years is the most we would ever fall below.'

'What would happen if you did fall below the minimum?'

'If we did that for two years in a row, then that would trigger a status review. People think that Wyoming and conservative wolf management are two things that don't go together. Yes, a lot of wolves are being killed in the predatory area, but that is because they are denning alongside livestock or conflicting with livestock, and that is damaging to people.

'In September 2014, a judge ruled that Wyoming's wolves would go back on the list and reinstated protections. I was on the ground then, and I saw us lose a lot of tolerance. I don't think that decision did anyone any favours. In fact, it swung a pendulum. I saw that more people wanted more wolves dead. I think we lost a decade of public tolerance that we had been nurturing – the tenor was so much more aggressive than it was before. It made me really sad – not because people liked wolves more before but because they had become more tolerant then. Also, because we need to trust and understand, and the groups that sued for the move to reinstate protections would not let us prove ourselves. The data is the data. We hold firm to the numbers – we have to hold firm and ask for trust, for people to be sure that we will do what we said we will do. We need patience from the public, from the pros and the antis.

'That whole act of trying to save the wolves just didn't work. We ended up growing a whole lot of wolves who were going to die because then the population became too big. It is a very complicated issue but something that Game and Fish has taken on. It is the responsibility we've been handed from the governor's office and from DC. We take it as a data-driven process, and we are working bottom-up to advise on legislation rather than using a top-down process. To me, this is not about politics. It's about the real data.

'On 6 June I will have a season setting meeting. I will stand up in front of the public and talk about it, and show what has happened to our wolves in the last year. I'll talk about the relationship between wolf numbers and breeding pairs and how that information relates to what we decide about hunting. I'll talk about the tight sideboards that we have to operate within. We don't just make this stuff up – we are rigorous, we track ninety collared wolves, and there is little margin for error in our work. That is an envious position to be in. It means we have good solid data that will drive the wolf hunting process and protect the wolves at the same time. But people will get emotional, and we will have those discussions. Yes, our management plan is agcentric because we have to address those issues of a ranching state.'

'Where do you get emotional?'

'My emotions come in as far as this really: we are really trying and I hate it that we have such a bad rap and a really bad name, but people don't always understand the reality. Living with wolves is the reality. Is it perfect? No. But we can manage it and make it work so that there can be a robust functioning population of wolves on our landscape and we can please hunters and ranchers. People are going to have to settle down about everything wolf so that we can get into a predictable, stable management routine over the years to come. I want people to give us the benefit of the doubt, to think "Ken Mills is a reasonable fella, he doesn't talk about what he feels, thinks or what should be done."

'This is the letter of the law, this is the data we have, this is how we have to do it. I don't like it. I couldn't kill a wolf, but I at least understand the playing field we are working within. If we opened up trapping and hunting seasons any more, we wouldn't meet our minimums, so we don't.

'You know, to my knowledge, no one has managed wolves in this way, using basic population theory and applying it to them. We are in a unique situation in Wyoming. I mean, how do you manage a relatively small population within a set of tight sideboards that no one else has? They have so many wolves that they can just keep trapping and shooting. I have a more mechanical approach – I am all about how to apply the theory and build a new type of management programme, to find out how the wolves function as a population. What I have learned is that we have to have adaptive management, which the court system hates because we can never tell them exactly what we are going to do.'

Ken sighs and looks a little perplexed and vacant.

'I think I have talked myself out,' he says. 'In fact, I can't remember when I have talked this much!'

We both laugh. He seems to have shocked himself.

'You know you were talking about C.S. Lewis earlier? Have you realised something? We are back in Narnia!'

'Yes! We came through the wardrobe.'

The snow has become more gentle again. I leave Ken in peace and wander around in the trees, searching for tracks while he figures out where to set up his cameras.

Hiking back down through the forest, I hear a weird noise like a distant engine and Ken stops dead, right in front of me.

'Did you hear that?'

'What was it?'

'Come and see.'

Quietly, we sneak through the snow, and Ken points at a sheltered patch underneath a group of trees. On a log, I can see the shape of a grouse.

'He made that noise?'

'Yes, it's a ruffed grouse. Let's wait – he'll do it again.'

We wait silently, standing side by side. The afternoon is growing late, the snow has soaked through my boots now, and my feet are cold and wet, as is my hair. It seems to take an age before the grouse moves again. He wriggles a little, changing his position, planting his feet firmly on the logs. Then he raises his wings to an odd angle.

'Here we go,' Ken whispers.

Clearly with his wings at that angle, the grouse is not going to take off. Very slowly, he claps, making a little thudding noise. Little by little, the 'clapping' movement speeds up. The sound he makes is just like a lawnmower that hasn't been started all winter, slow at first and then turning over with more and more speed and excitement until … it stops suddenly, and the grouse just sits there.

'He's listening.'

'What for?'

'He's staking his territory and wondering if there are any other grouse around.'

'Including lady grouse.'

'Oh, yes. That is why they are called "fool hands".'

'Seems like a funny way to catch a lady.'

I am relieved to see the little 4x4 perched where we left it, and I'm happy to get sheltered inside. The ride down the mountain is as fast, sketchy, slippery and hilarious as the ride up, but finally, we are back in Ken's truck with the heating on. As we ride back over the mountain pass, Ken tells me how much he loves his job.

'I love the mix of it,' he says. 'The analysis, the horse riding, the challenges, gathering data. Wolves are a difficult species. Everything surrounding them is difficult, but I get to be a part of implementing how they are managed and proposing decisions that will keep them robust. And I get to talk to the public about it. I get to make a difference. Today I'm up a mountain. Tomorrow I'm in first grade at school for my annual talk.'

That's Ken. People and wolves – a pretty difficult combo. No wonder he doesn't talk much.

CHAPTER TWENTY-FOUR

At one point we lost the most calves in the lower 48

In the fall of 1996 strange things were happening on the Diamond G Ranch in Dubois. As the crow flies, the ranch is only around twenty-five miles from Yellowstone. After a thirty-minute drive along slippery mud tracks, I am finally drinking coffee and eating cheese sandwiches in the Diamond G kitchen with Jon and Deb, who have managed all 48,000 acres for the last thirty years.

'It started when we found that heifer in the creek, didn't it?' Deb asks Jon.

'Oh, yes, it did. We took one look at her drowned in the creek and—'

'We asked ourselves why on earth would she have got herself into the creek and drowned in there?'

'I mean, she could have gotten out?'

'She could if she had wanted to. That was the start of it really.'

Jon and Deb have been together since fifth grade at school when Jon joined Mountain View school in Casper, Wyoming. Of course, they were just great friends then, but apparently, Jon had gone home from school and told his mum that Deb was the girl he would marry. It wasn't until after high school that they met again at the rodeo.

'I did pole bending and barrels and team roping,' says Deb.

'She was the first girl to get into the state finals for roping – that was an ice-breaker back then,' says Jon proudly. 'Now the girls are kicking all their butts.'

After so many years of living in the remote wilderness and ranching together, Jon and Deb clearly know exactly what the other is thinking. They finish each other's sentences or fill in the gaps while regarding each other lovingly.

The couple wanders in and out of an office to one side of the kitchen as we chat. They bring marked maps, piles of binders, paperwork and pictures.

'After that,' Deb says, 'we started getting more suspicious, until finally—'

'We pulled over the slide gate on five mile up there—'

'And I was on my horse,' says Deb, 'and I said, "Jon, there's something wrong with that rock," and then two ears appeared and a wolf stood up. It wasn't a rock at all. Wolves are curious, not confrontational.'

'So how did that play out with the cattle? When did you first realise you had a problem?'

'Ah, that night?' Deb looks at John.

'You see,' interjects John, 'we sleep with the bedroom windows open, and we heard the cattle.'

'It was a dreadful noise, coming from the cow-calf herd,' adds Deb with a shake of her head. 'So we grabbed our clothes and hauled on down there in the Land Rover. And when we turned on the spots, there were the wolves chasing the cattle.'

'What did you do?' I ask.

'We managed to chase them off that night, but it didn't stop them from coming back.'

'That must have been a pretty massive shock.'

'No,' Jon says, 'we anticipated this. When we knew they were planning to reintroduce wolves, we spent a lot of time researching. We knew it was just a matter of time before they showed up. Those wolves were unsettled when they were released, so they were maybe more random in their dispersal.'

I catch the sweet scent of the vase of daffodils placed on the breakfast bar as Deb hands me an A4 photo of two black wolves standing in long grass and staring at the camera.

'That's them. That's how tame they were. I could get really close to them, and they weren't bothered.'

'Deb is a photographer and she didn't even have a long lens back there, did you?'

'Nope.'

'Doesn't look like you needed one,' I add.

'The bigger one there is the male. Carter's Hope, he was called. He was one of the ones caught in Canada, and Carter was the name of the guy who caught him. That's his female. It was just the pair of them at first.'

The whole of the counter is covered in binders and photos now.

'You seem to have immersed yourself in the subject?'

'We sure did,' Jon says. 'We have been in a unique position, so we go round the country giving talks. We tell them how it is. It is about the impact, not our feelings. We had already been dealing with a high intensity of grizzly bears for years. I mean, we had forty-two bears in this area that no one knew existed. They were new to the population when we reported them. We helped with research for Game and Fish and Wildlife Services, and we helped research a book with the Craigheads. Where is that book, Deb?'

Deb hands me a huge volume called *The Grizzly Bears of Yellowstone 1959–1992*.

'We were working out how to live with the bears,' Jon continues, 'how to manage a ranch with so many bears on the property, what to do with our carcasses, and all of that.'

'I guess we were used to it, we just did the same for wolves,' says Deb. 'Was it that first litter in '97 that we got a collar on Tracy?'

'I think so. We named all the wolves, you see. Was that Tracy that got killed by a mule?'

Jon is busy flicking his way through binder after binder, photo after photo.

'I think it was Tracy. It's all written down in there somewhere, Jon.'

'Oh look, here's Bradley,' he says, pulling out a photo of a huge grey wolf asleep in the arms of three smiling teenage girls, his tongue lolling out of the corner of his mouth. It is as much as the girls can do to hold him – he is longer than the three of them are standing side by side.

'I'm guessing the reason he was so fast asleep was so that you could collar him?'

'He was, yes. We have to trap them. We use teeth traps – they are better because they don't cut off the circulation like the ones without teeth – and Game and Fish, or whoever we are working with will check the trap at least once a day. They take care of the collaring, but we have the frequencies so we can monitor where they are going.'

I can hardly take my eyes off the picture of the wolf. He is so impressive, so utterly beautiful.

'That is a big wolf.'

'Bradley? He wasn't so big, you know. What was he, Deb?'

'Around 120 pounds, I think. We caught one wolf that was 145 pounds. That was Ron, wasn't it?'

'No, not Ron, another one but with a similar name.'

'Right, what was that name?'

'It'll come back to us. Anyhow, that first pack we had in '97, that was our first year of wolf depredation. We lost sixty-one calves. I didn't know it but, when we brought the cows and calves up in the spring, I turned 800 pairs out right on top of a wolf den.'

Deb flicks through to some gruesome photos.

'This is a calf, see? This is what we have to do.' The photo she points at shows a calf's head with its skin peeled back. 'This is the only way you can see the wounds.' She points out the black puncture wounds where the wolves' teeth have injured the calf. 'The wolves' teeth are not that sharp, so often you can't see anything on the outside, but the bleeding has all happened on the inside from the pressure.'

'Sometimes,' Jon adds, 'there is nothing to see at all and, if you didn't know, it would be impossible to tell how that animal died.'

'Do they hamstring them?' I ask.

'We very seldom have seen that, although they can do it – but normally with elk rather than cattle. We have seen them attack an animal, pulverise it and come back two to three days later and kill it then.'

'That sounds pretty cruel,' I wince. 'I guess that's why so many people hate them, for that and surplus killing.'

'Look, you have to understand that wolves kill out of necessity. We haven't seen them kill for "sport". If they are teaching the pups to hunt in the fall, then they may surplus-kill, but we have never seen a mass slaughter. And we would have seen it by now if they were doing that. This is the way wolves behave, this is what they do. They aren't humans.'

'How did you feel about it?'

'We don't feel either way. I mean, it was a lot of extra work. We had to ride all around the allotment every day to find carcasses. And of course, it took us a while to work out what was going on.'

'This map shows the kills in '97.'

Deb props a map up against the fridge. It shows the land they look after, and she pulls a transparent overlay on top that is covered in wolf head symbols. Right in the middle, just beside the ranch, is the wolf den. Around the den are many, many red wolf heads with dates and numbers alongside them.

'These red heads represent just some of the kills from that year. Do you see how many there are?'

It is impossible to count.

'We recorded most of them on this map to give people a sense of wolf behaviour on a ranch,' Jon says. 'We also lost a lot of elk here that first season. We would often count 600 elk here on a spring day, just eating our grass, till around July time when they would move on. We worked out that elk cost us around $48,000 a year in grass lost. The most we ever had was 2,200 here on the ranch. I have that information on a graph in here somewhere. We lost 400 head of elk in the first couple of years, and we did elk studies with Game and Fish. Here it is,' Jon says, showing me a diagram illustrating how wolves affected the population of elk. 'They change the way the elk behave. Before the wolves, they would spread themselves all over the ranch, just lazing around and enjoying our grazing, and their migration would pass slowly through the middle of us. With the wolves around, they'll bunch up tighter and move on faster.'

'And actually,' says Deb, 'they change their route quite a bit. They'll travel higher, so they don't even come down this low on the ranch.'

'Did you continue to lose cattle the following year?'

'We did because in '97 they had five pups. That following year we lost fifty-six calves, and at that point, we were the most chronic ranch in the lower 48 due to depredation.'

'It must have been really hard. I mean, did you hate them?'

'Me, not so much,' says Jon, looking at his wife, 'but Deb did.'

'It is really hard,' she says. 'And the hardest of all for me was when they started taking our dogs. Then it was personal for me. One old dog we had, she was fourteen and stiff, and we let her out. We didn't know wolves were around then, but she must have because she went around the back and we never saw her again. We couldn't find her that night, but the next morning we did – they ate part of her and dragged the rest of her out the back a little ways. Sometimes, as far as I'm concerned, they should all be hanging on the wall.'

Jon is flicking through a binder as Deb talks. He pushes it over to me and points to an image. I see really deep bite wounds in long white fur.

'What is it?'

'That was our Great Pyrenees – we had two for herding and guarding dogs. The wolves took her right there on the front porch. Just ripped her apart.'

I look at Deb. She is shaking her head again and looks as though her eyes will fill with tears at any point.

'Didn't you want to leave?' I ask. 'I mean, you must have thought about it. I'm sure I would.'

'No, never,' she says. 'We've been here thirty years. We never wanted to move, but we often would get frustrated at the way the wolves were managed.'

'The dogs were really close to us,' Jon says, 'but you have to understand that wolves will do what wolves will do. So I had to think, "what mistake did I make? How can I change my behaviour?" So we built that big kennel out the back, and we haven't lost a dog since. You've also got to know what they are going to do at different times of the year – when they get close to breeding season, wolves will go after any canine in their area – coyotes or dogs. That's just what they do, and I hadn't protected our dogs.'

There are other photos.

'We lost that beautiful little colt, remember? That broke my heart,' Deb says.

'I remember. We've lost horses and colts to the wolves. This here is a calf that Deb patched up, and we did manage to save that one.'

The photo shows a calf in a stall. Three deep wounds run down his shoulder, with others less deep in between, and the subcutaneous fat bulging out of the deeper ones. It is a mess.

'These guys weren't hamstringing. We just don't see that – they'll go after the shoulder and the nose first. We used to have registered Red Angus – this one was a beauty. She did get attacked horribly by a grizzly bear, but we pulled her through, and she lived for years afterwards.'

Jon tells me a cow like that is worth $8–10,000. Now I know what ranchers mean when they talk about financial loss.

'So what happened to the original pair?'

'Carter's Hope is in the museum now. They shot him in '97 or '98. We asked them not to. We wanted to see if we could live with them, but he was killing a lot of cattle, and it was out of our hands. The day before, Game and Fish told us what they were going to do, and they told us not to tell anyone. That night we got death threats.'

'That was scary,' Deb says. 'They told us that they were going to kill our grandkids and gave us their names and addresses to prove they knew where our family was.'

'I can't imagine what that was like. And you still didn't think of moving.'

'It was horrible, but the death threats weren't going to affect us and make us leave a place we love.'

'You've got to have a backbone somewhere,' Jon says, standing more upright by the fridge. I get a clear view of the guts and determination in his eyes. 'It was interesting – when the wolf family was all here, we had less depredation. When they shot the adults and left the pups as yearlings, we had more. You see, those pups didn't know how to hunt elk yet, and most of the elk had migrated out, so the cattle were an easy opportunity if they were going to stay alive.'

It wasn't just the pro-wolfers that were threatening Jon and Deb. He tells me he's not really very popular with other ranchers either.

'A couple of times, I've been threatened for being pro-wolf because other ranchers have a different agenda when it comes to wolves and bears. As the manager here, I lose cattle, I make the decision, learn how to ranch with them, and take the blame or blame the predators for what they do naturally. Management means killing whether you like that word or not. Wolves were brought back to rehabilitate Yellowstone, and they did a wonderful job.

'This is a people issue, not a wildlife issue. Everyone has a different agenda. The hunters and the outfitters all want elk, and they want to blame the wolves rather than themselves for taking the elk numbers down. Think about it: a wolf eats 1.8 elk per month per wolf; the average age of an elk eaten is twelve years old; however, 50 percent of a wolf's diet of elk consists of calves. That means we are losing our recruitments, sure. So what is the magic number for elk? We have to work it out.

'I'm not a wolf hater, but wolves shouldn't have been brought back to Yellowstone. It just isn't big enough, and everything around it is domestic livestock and hunting, so most of Wyoming falls into the predator zone. There are problems everywhere, depending on people's agendas, but if everyone could take a few steps to the centre, we could solve all of it. We have to look to compromise: how do we make it work for both humans and wildlife? If we get good packs – the ones that aren't on our cattle – we shouldn't kill them. Until we can come to the middle and talk over the table, we can't come to an agreement on any of it. Wolves could be anywhere in Wyoming, but unless they are a problem, they should be left.

'If it is happening to you, then it is significant – you will lose money to this. We've allowed them to get out of hand, yes, but then we have to consider that this is the Wild West. You live here, you deal with it. It comes down to attitude and how you live with wolves. That is up to the landowner – do you accept them, or do you tolerate them? If you tolerate them, you put them in a box. If you accept them for what they are – social, pretty predictable – if you spend time with them and watch them, then you know what you

are talking about. We have seen so many things just by
being here and observing. Remember those five pups we
found, Deb?'

'Ha! There was no sneaking up on them, that's for sure.
They were on their own. They were about two months
old then, and the adults were off hunting, but they were
lying in a square. Each one had his nose on the next one's
flank, and one in the middle, so you couldn't approach them
from any angle without one of them noticing you.'

'Wildlife, to me, is what it is all about,' says John. 'The
more you know about something, the better off you are.
If you want to find a carcass, just look for the ravens – they
will tell you where they are. Finding carcasses is hard. No
one wants to look at an animal they have raised and find
it dead because of a predator, but we have to remember
we are living in their backyard too. And we have to decide
what is our priority: economy? The elk herd? The environment? Agriculture?'

'Do you remember what the judge said?' Deb asks Jon.

'Well, our boss tried to sue the government over the
wolves, but he lost. What did he say, that judge?'

'He said the problem was that the wolf introduction
hadn't been in place long enough,' Deb replies.

'That's right, but what we saw was that the longer the elk
and cows were exposed to wolves, the more their behaviour
changed, and they were more aware of self-defence. They
began to move as a unit, keeping tight together instead of
spreading out. So the last year we had the wolves with the
cow-calf operation we had zero predation, and that had
been playing out over the years. In 1997 we lost sixty-one
animals. In 1998 it was fifty-six, then in 1999, it was fifty-
three. As far as I'm concerned, you have to understand how
the animal works, then you can understand how to ranch
with them. We had already worked with bears, so we knew
that. We didn't want them shot. Instead of saying "let's shoot
it," let's ask why they are here.

'We have had twenty-seven dens in that drainage and, when you think about it, what they are doing is getting ahead of the elk migration. They move in, den and have pups, and then they are ready for the elk to arrive and calve. After the wolves moved in and established themselves, the elk changed their pattern. Over the years, we were given shoot-on-sight permits.'

'Do you have wolves here now?'

'No, there just aren't enough. Game and Fish killed as many as they could before relisting, and so now they find themselves short on wolves. It's complicated in Wyoming to get the numbers right between the predator zone, the trophy game zone and Yellowstone. And of course, there is a lot of civil disobedience, a lot of extremists.'

'By "civil disobedience" do you mean wolf shooting?'

'Right. There is a lot of shooting going on, from what I hear. These people talk about it, and we hear about it. They see bears and wolves as competition. There is so much misinformation out there that it polarises people. There is no one in the middle any more – there never was. You need to understand the animal. It depends on your expectations, and everyone took a side before anything ever happened.

'That fourth pack that we had here was great. The wolves weren't interested in cattle – they would walk through the herd of cattle to get to the elk, and we had no problem with them. I was happy to have them around, but they vary.

'It's odd not to have wolves. This is the first year we haven't had any when there used to be so many. The Washiki pack that was started by Carter's Hope – they were all black, but one – they got up to twenty-three. Half of them went and formed the East Fork pack, then some more to the Wiggins Fork pack. One day, when we were looking at the frequencies we had on all the collars, we realised we had five packs here on the ranch. We would have cow camps here, and we had three or four range riders on the mountain at all times. It takes a long time to ride around 48,000 useable acres – all day.'

'Was the range riding useful? Did you try the usual anti-wolf measures?'

'None of it was really effective for us. We tried all this crap – flags on the fences and all of that. These animals are clever. I mean, what would you do if you were a predator and you saw some people coming in on horses? They would just hide in the woods and wait for us to go by, then get on with their business. If we were around too much in the day, they would just turn nocturnal, but if we ever got too close to a den or rendezvous sight, then you would find yourself with a wolf on either side of you and behind you. They would get close, and they would gently steer us away from the sight, but they didn't move their site, they just moved us. Predators are forgiving, flexible and adaptable. Wolves and bears aren't just wanton killers.'

'Do you remember that bear?' Deb asks Jon, laughing. 'We had this bear that they were trying to get a collar on, so we had a trap out, and we had cameras on it so we could see what he was doing because we just couldn't get him. He was bringing a stick and going up right behind the trap, then he'd put the stick in the trap to set it off and take the bait. He was really clever.'

'Wolves do the same thing all the time. Once you have figured out what their pattern is, you can deal with them. If you have a good pack, you don't want to take them out. A few years ago we had a great pack, and Game and Fish wanted to take them out for predation on another ranch, but with collars, I could prove that it wasn't our wolves. We could have ended up losing a good pack and gaining a bad one. That's why I'm not a wolf hater. I always ask myself what can I do differently.'

Deb is still flicking through the photo album.

'Do you remember this day, Jon? It was 2016 and Jon and I were riding out together, which we didn't get the time for that often in those days.'

'It was normally just Deb. She'd spend all day on that horse – she had to.'

'But on this day we were together. We saw elk in the distance, and we could hear elk talking to each other – you know when they sound like they aren't very happy. The horses were a little bit nervous too. There was a deep gully right in front of us and suddenly out popped this big black wolf. If that had been a bear, we'd have been too close, but the wolf ran off, so we went to see what all the noise was about. He had just taken an elk down. He'd gone for his nose and shoulders first – but look here, you can see that his whole body is covered in bite marks.'

It was true, the elk was sitting at the bottom of the gully with his head raised. There was no blood to be seen, but puncture marks – looking a bit like cigarette burns – were all over his rump and shoulders. He was still alive while the wolf was trying to eat him.

'But how do you blame 'em for that? The wolf is just doing what wolves do,' says Jon. 'Do I want to watch it? No. Do I like it? No. But that is what they have to do to make a living. They will kill most things, so often the things we have witnessed on this ranch have put people's research to the test. What they say is "abnormal behaviour" we have seen is normal. I mean wolves sunbathing on your front lawn – "abnormal" but normal around here. They are curious, and they will come and follow us. They don't attack, but they want to know what we are up to – abnormal? No, normal round here. See this heifer?' Jon shows me another picture and points to a blob of black fur and blood next to the heifer. 'What do you think that is? It's a bear cub. They killed the heifer and then left, then they saw a grizzly sow and her two cubs come to eat it, so they came back and killed the cubs.'

'We've had a big 350-pound grizzly bear killed by the wolves,' adds Deb.

'We had one researcher say, "Just go outside with a recording of helicopter noises. Those wolves hate helicopters – they know they are going to get darted if a helicopter comes by."' Both Deb and Jon are laughing now, tears filling their eyes. 'So I'm out there, wandering around like an idiot with a tape player thing, and do you know what those wolves did? They just came over the hill and watched us. They weren't frightened in the least bit, they were just curious.'

'Then, who was that other guy? The one from Alaska,' says Deb, laughing.

'Right, he came down to help us out, and we were sitting on the front porch talking, and the wolves just trotted up the road right in front of us while we were having a beer. Mike Jimenez had just written his theses on how wolves do not use roads, and then he came here and started laughing and said that blew his theory out of the water.'

'You've put up with a lot more than most would, though,' I say.

'Two governors have been elected because of what they say they are going to do about wolves – Mead and the one before him,' says Jon. 'We went to Congress and testified. We wanted to know how are we going to make this work. What did they say? "We will help you after the election." No one could really believe that we wanted to preserve the pack, that we wanted to try to live with them like we did with the bears. With them, we learned pretty quickly that if they are using a certain part of the allotment – like in the fall when they are gathering berries – then we don't graze our cattle there. It works. I have friends in Canada, in the lower provinces, who live with wolves and their cattle don't get taken. How do they make it work? And we need to get our numbers, right? How many is enough? We are losing our population right now. All management of wildlife seems to be after the fact. Wolf packs are changing the rules, and we are on the ground observing it all – wolf packs changing the elk migration routes – all of it.'

'Right now,' says Deb, 'because we have no wolves, the elk are migrating lower again. Everything is a cycle, and the wolves are being overheated again.'

'And so much of it comes down to how you ranch. You can't moan about predators if you leave cow carcasses lying around and give them a taste of beef.'

'Remember that one you buried right up there?' asks Deb, turning to me she says, 'Jon buries our carcasses really deep to stop them from being dug up, but what was that one? A year-old?'

'Yes, maybe more.'

'Well, a badger starting digging on that spot, and then the smells must have come up because one day I could see a bear digging up there. She was already in up to her shoulders, and I could see she had a cub with her. She dug all the way down and then the wolves came and chased her off. She came belting down by the house, right by the kitchen window where I was washing up, and the wolves were still following her. I thought they weren't going to give up, but she did get away and then later she came back for her cub.'

'I still find it incredible – after all that you have been through – that you haven't fallen to hatred.'

'There are other people like us, but they are scared to say anything,' says Jon. 'If you didn't have compassion for what you live with, then you would be screwed. How miserable would we be if we couldn't accept it? It would be a hate relationship. It is a lot easier to blend in and be part of the environment. We live that – we are that. Otherwise, we would live in hate and misery. You have to give space. You have to understand all the species and how they fit together. We've lived with wolves for twenty years now, and for the first time, there aren't any on the ranch. Something's got to be wrong.'

Jon and Deb have lived the consequences of the wolf introduction for over twenty years. When they wave me off, I am filled with awe for their attitude, their tolerance and bravery. Faced with their situation, even I – a person who is so passionate about the wolves – am not sure if I could have been the same way.

CHAPTER TWENTY-FIVE

Elk for breakfast

Early one morning, I decide to take a trip down to Buffalo Valley, the place where my journey began. This time I know I will not hear the Phantom Springs pack – they are no longer here. The only thing that seems guaranteed for wolves is change.

In 2016 the Phantom Springs pack were replaced by the Lower Gros Ventre pack, who used to hang out in Moose and on the Gros Ventre buttes and even on the lower side lake, which is quite some territory, and historically they had conflicts with livestock.

For whatever reason, after an unfortunate few years, the Phantom Springs pack failed to reproduce that year, and so they were down to just four wolves. That was the year the wolf population were facing challenges, and their numbers were up – literally. A higher population meant that there was more disease and wolf conflict than had ever been seen in the valley, and the elk had redistributed themselves, which only added to the changes. The Lower Gros Ventre pack knew the Phantom Springs pack were struggling and, at only four wolves, could not very well defend themselves. So they invaded, they killed the alpha female and took over their territory, partly because they were being squeezed out of their own lands by the Pinnacle Peak pack.

This morning I am in luck. I drive by Tracy Lake, and suddenly on the right, close to the road, I see them.

I stop the car a distance away. They have killed an elk. I count six grey wolves. Four are feeding, and two are standing at a distance.

There is no one else around. No wolf watchers, no rancher, no hunter, no biologist. I have only my own thoughts to listen to as I watch. I am, once again, filled with a familiar sense of wonder. Despite all I have learned, that hasn't gone away. I feel so privileged to watch these wolves in private, to know that they are still going about their business in the wilderness whether I am there or not, whatever anyone thinks about them.

What is the truth about wolves? I guess that is what I set out to discover. This is it: them, in front of me, feeding as a family. This is the truth about wolves.

Yet, on the trail of them, I have found out more about people than I have wolves. I have found out about myself too – mainly that I love listening to people, especially those who see things differently to me. In fact, I haven't met one person I didn't like, even though I have been seeking people out who will have an opposite view to me. The truth is that we all have something to say. I have not met one person who has no opinion about wolves. But the problem with wolves is not them – it is us. We have stopped listening to each other.

This isn't as simple as love and hate, although I still love wolves, I still love Wyoming, I still love the Wild West and the wilderness. If anything, I love it even more now – I now have the addiction. I want to carry on finding out about wolves. My curiosity is not sated. I want to know more about what makes us so different from them. I want to be part of the pack. I want to feel a sense of loyalty, of freedom, that family is all I have to fight for and that if it is my time, it is my time – a sense that nothing else really matters. Wolves are stripped down, basic, and I'm curious to know how that feels. I'm not sure we really need anything else.

As a mother, I know the feeling of being denned up with cubs, of being protected by my pack. It is instinctive and the best feeling I know. Moving to a different country means that I no longer have my pack around me, and that has been hard. Our families are disparate in our modern society. Perhaps we miss those bonds, maybe we have forgotten how to live as a tribe. We have family, grief and struggle in common with wolves. As humans, we may be tempted to believe our lives are worth more than theirs because of religion, because we have art, we have technology – we have many reasons. But none of them gives us the right to believe that.

Do we love wolves because biologically speaking, they are so close to who we are? Because we identify with them? And do we fear and hate them because we recognise cruelty in them that we also have within us that we don't want to face? I realise my own fascination has partly been not about whether we can live with wolves, but about how we have forgotten how to live like them: to be in the moment, to have a clear set of 'values' even if they are only biological, to play, to live and die with gusto because there is no other choice, to be comfortable in the natural world.

The wolves in front of me are no longer so focused on the kill. I guess they are full. They have finished feeding for the morning, but they may come back later for another snack.

They look over, seeing my car, and they obviously know they are too close to the road to collapse in post-breakfast repose. One by one they turn around and trot off, heading for a distant patch of trees. One by one, as they reach the rise in the green Wyoming grassland, they turn to look back. Are they looking at the elk or me? I can't quite tell.

The last grey wolf disappears, melting into the trees, and I switch my engine back on. The wolves have left.

I remain passionate and compassionate. I'm hopeful that the last grey wolf will never howl his loneliness to mankind here again, and I understand so much more than I did before I came here. A wolf's howl is built of human things. The musical score of their wild lives stirs stories of loss and heartache and of community in our souls – things we identify with as social animals if we are brave enough. I have not met one person on my journey who hasn't been moved by it.

Love 'em or hate 'em, we are at the beginning of a new chapter of wolf history and, once again, the wolves will not get to write it alone. Humans will edit their story. Humans who often will have very different narratives.

I drive by the precipice, slowing down for grouse corner, remembering the grouse frozen with fear that made us laugh, and all the school runs up and down this road discovering widlife. The voices of all the people I have spoken to echo in my head, each telling their version of the story, but the one who speaks loudest is Doris Platts: 'Draw your own inferences and conclusions from the accounts here gathered.'

It is together that we must try to write the next chapters for the wolves.

Acknowledgements

No one will deny that writing a book is hard, and although I got the words onto the page, this book took a team, and I wouldn't have it any other way.

Usually, a book comes from a burning desire to say something, but it wasn't really my voice that was important to me, it was the voices of the people I discovered when I moved to the US. I realised that our lives, our politics, even our social media pages are so much about what we have to say that we have forgotten how to really listen to what others have to say – how to listen to understand. In this case, to understand why so many perceive the same incredible wolf in different ways. Passions run high when it comes to wolves – everyone has something to say, but it isn't always something good.

So my debt of thanks goes first to the people who took a risk and understood that I wanted to listen to understand, rather than to judge.

Verne, the ex-trophy hunter who showed me so much kindness and other hunters who spoke with me at length including Ed Bangs and Ron Hoag, Malou and her family who hosted me and talked so openly. The wolf watchers of Yellowstone who always share, especially Wolf Man Cliff, Jan of Beds and Buns in Cooke City, Lisa Robertson, Steve Cain and Kira Cassidy who were beyond generous with their time and knowledge. Ken Mills and Game and Fish for allowing me an insight into the challenges they face to keep our wolf population in Wyoming. Jon and Deb at Diamond G ranch for sharing the details of a life ranching with predators.

Julie Bailey at Bloomsbury – thank you for being so excited and supportive of this idea – you have been a joy to work with.

The huge team at Bloomsbury who have been behind me all the way.

Sophie and Derek Craighead and Noa and Ted Staryk for their ongoing support.

Ronan Donovan who shared his photo for our cover with such excitement.

Then there were the people who gave me faith in myself when I had none left.

Julie Elledge, and especially my agent Gill McLay — without whom I would not have had the courage to keep putting pen to paper and believe in what I was writing. Tina Price my PR without whose constant love and support I would still be stuck under a TV in Ealing.

My family — Fred, Gus and Arthur, you have watched me fail and fall and try for so long — such is the writer's way, and every time I get back up, it is because of you. I hope I have made you proud.

Charlie — I did it.

Bibliography and Further Reading

Askins, Renee. Shadow Mountain: A Memoir of Wolves, a Woman, and the
    Wild (Achor Books, 2002).
After forming an intense bond with a wolf cub she raised as part of her
undergraduate research, Renee Askins was inspired to found the Wolf
Fund. As head of this organisation, she made it her goal to restore wolves
to Yellowstone National Park.

Dutcher, Jim. *Wolves at Our Door: The Extraordinary Couple Who Lived with
    Wolves* (Touchstone, 2003).
Determined to overcome misconceptions of wolves, Jim and Jamie
Dutcher spent six years in a tented camp on the edge of Idaho's wilderness,
living with and filming a pack of wolves.

Eisenberg, Cristina. *The Wolf's Tooth – Keystone Predators, Trophic Cascades,
    and Biodiversity* (Island Press, 2010).
An ecology textbook written in the first person.

Jans, Nic. *A Wolf Called Romeo* (Virgin Books, 2014).
The unlikely true story of a six-year friendship between a wild, oddly
gentle black wolf and the people and dogs of Juneau, Alaska.

Lopez, Barry. *Of Wolves and Men* (Simon & Schuster, 1979).
A classic exploration of humanity's complicated relationship with and
understanding of wolves.

Mech, L. David and Boitani, Luigi. *Wolves: Behavior, Ecology and Conservation*
    (University of Chicago Press, 2007).
A systematic and comprehensive overview of wolf biology.

Mowat, Farley. *Never Cry Wolf* (Little, Brown, 1996).
A lighthearted, pro-wolf book written by a biologist who encountered
wolves while studying whether or not they were responsible for declining
numbers of caribou.

Peterson, Brenda. *Wolf Nation – The Life, Death and Return of Wild American
    Wolves* (Da Capo Press, 2017).
Peterson blends science, history and memoir to dramatise the epic battle to
restore wolves and with them, the landscape and ecology of the continent.

Smith, Douglas and Ferguson, Gary. *Decade of the Wolf* (The Lyons Press, 2006)
Doug Smith was and continues to be the leader of the wolf reintroduction project in Yellowstone. This book covers the story of the controversial reintroduction and the years that followed.

Thayer, Helen. *Three Among the Wolves* (Sasquatch Books, 2004).
A true-life adventure tale combined with a fascinating natural history of the wolf.

Websites

Faces of the Endangered
This is the Facebook page of 13-year-old wildlife artist and animal advocate, Bria Neff, who has donated over $65,000 by selling and displaying her paintings of vulnerable animals and landscapes.
facebook.com/Facesoftheendangered/

Wolf Conservation Centre
nywolf.org

Index